World Water Resources

Volume 1

Series Editor
Vijay P. Singh, Texas A&M University, College Station, TX, USA

This series aims to publish books, monographs and contributed volumes on water resources in the world, with particular focus per volume on water resources of a particular country or region. With the freshwater supplies becoming an increasingly important and scarce commodity, it is important to have under one cover up to date literature published on water resources and their management, e.g. lessons learnt or details from one river basin may be quite useful for other basins. Also, it is important that national and international river basins are managed, keeping each country's interest and environment in mind. The need for dialog is being heightened by climate change and global warming. It is hoped that the Series will make a contribution to this dialog. The volumes in the series ideally would follow a "Three Part" approach as outlined below: In the chapters in the first Part *Sources of Freshwater* would be covered, like water resources of river basins; water resources of lake basins, including surface water and under river flow; groundwater; desalination; and snow cover/ice caps. In the second Part the chapters would include topics like: *Water Use and Consumption*, e.g. irrigation, industrial, domestic, recreational etc. In the third Part in different chapters more miscellaneous items can be covered like impacts of anthropogenic effects on water resources; impact of global warning and climate change on water resources; river basin management; river compacts and treaties; lake basin management; national development and water resources management; peace and water resources; economics of water resources development; water resources and civilization; politics and water resources; water-energy-food nexus; water security and sustainability; large water resources projects; ancient water works; and challenges for the future. Authored and edited volumes are welcomed to the series. Editor or co-editors would solicit colleagues to write chapters that make up the edited book. For an edited book, it is anticipated that there would be about 12–15 chapters in a book of about 300 pages. Books in the Series could also be authored by one person or several co-authors without inviting others to prepare separate chapters. The volumes in the Series would tend to follow the "Three Part" approach as outlined above. Topics that are of current interest can be added as well.

Readership

Readers would be university researchers, governmental agencies, NGOs, research institutes, and industry. It is also envisaged that conservation groups and those interested in water resources management would find some of the books of great interest. Comments or suggestions for future volumes are welcomed.

Series Editor:

Vijay P. Singh, Department of Biological and Agricultural Engineering & Zachry Department of Civil Engineering, Texas A & M University, USA, vsingh@tamu.edu

More information about this series at http://www.springer.com/series/15410

Elias Salameh · Musa Shteiwi
Marwan Al Raggad

Water Resources of Jordan

Political, Social and Economic Implications
of Scarce Water Resources

 Springer

Elias Salameh
Center for Strategic Studies,
University of Jordan
Amman, Jordan

Musa Shteiwi
Center for Strategic Studies,
University of Jordan
Amman, Jordan

Marwan Al Raggad
Water, Energy and Environment Center
University of Jordan
Amman, Jordan

ISSN 2509-7385   ISSN 2509-7393   (electronic)
World Water Resources
ISBN 978-3-030-08528-5   ISBN 978-3-319-77748-1   (eBook)
https://doi.org/10.1007/978-3-319-77748-1

Preface

One of the major challenges facing Jordan is its severe shortage of water resources. As one of the most water poor countries in the world, water scarcity is the norm. This scarcity in Jordan is compounded by strategy- and policy-related developments and social factors: the most significant being the rapid rise in population. The population of Jordan has increased tenfold since the 1950s. In addition to the high rate of population growth, Jordan has been subjected to a series of massive refugee influxes since 1948, most important of which are the influx of Palestinian refugees over the past decades and the current wave of Syrian refugees with more than one million Syrian refugees now residing in Jordan.

This book discusses the need for a regional approach to solving the problem of water scarcity not only in Jordan but also in other countries in the region. The book ends with some practical recommendations on how to deal with the water problem in Jordan.

Furthermore, over the last few decades Jordan's water resources have also been continuously exposed to rapid degradation, not only because of active pollution introduced by liquid or solid wastes, but also, and increasingly, by passive degradation due to salinization as a result of the over-pumping and depletion of the groundwater resources base. Widely applied remedial measures during the last decade have alleviated water catastrophes and the inability of the country to provide water of sufficient quantity and quality for human life and subsistence.

This book is designed to provide an overview of the water situation in Jordan and how it has been affected by the last few decades of rapid socioeconomic development. For this purpose, the first chapters describe the availability of water resources in the country.

The section on water quality provides information about the original water qualities in the different regions of the country and how they have been affected by pollution such as that caused by cesspools, treated and untreated waste water, industrial waste water, solid wastes, irrigation return flows, salt water intrusions, and the upcoming of salt water bodies.

Following this, the loss of resources, declines or losses of water production facilities, and water quality degradation as a result of population growth are discussed. Future projects to develop additional resources to substitute degraded resources and increase water availability for the use of coming generations are put forward. The book also touches on the issue of social cost; the cost incurred by Jordanian society as a result of water pollution and depletion.

The book also discusses the managerial, technological, and pricing policies the country is envisaging to achieve a sustainable water resources base taking into consideration intergeneration equities in terms of quality degradation and overexploitation limiting factors.

Amman, Jordan   Elias Salameh
                                                                        Musa Shteiwi
                                                                   Marwan Al Raggad

Acknowledgments

The authors wish to thank the many people who have helped in the preparation of this manuscript. The Ministry of Water and Irrigation kindly provided data on water resources, uses, and valuable information about past and current projects of the ministry. Special thanks go to the ministry staff particularly to Eng. Thair Al-Momani for his valuable cooperation.

We also extend our warm thanks to the administrative staff at the Center for Strategic Studies at the University of Jordan for their continuous support throughout the preparation of this research and manuscript over the last 6 months.

The authors are highly indebted to Dr. Ghaida Abdallat for critically reading, commenting, and suggesting improvements to the book.

Thanks also go to the Federal Ministry of Education and Research (BMBF), Germany, and the German Research Foundation: Deutsche Forschungsgemeinschaft (DFG) served as a source of information on the results of various projects of both institutions.

Here we gratefully acknowledge their support in our research on the development of the water sector in Jordan.

Contents

About the Authors

Dr. Elias Salameh is a professor of hydrogeology and hydrochemistry at the University of Jordan. He obtained his Doctor of Science degree from the Technical University of Munich, Germany. He founded the Water Research and Study Center at the University of Jordan and served as its director from 1983 to 1992. From 2004 to 2005, Professor Salameh served as the chairman of the Founding Committee of the German/Jordanian University. He has served as a member of the Royal Committee on Water since its formation in 2007. He also served as a member of the Board of Trustees of Balqa Applied University from 2008 to 2015. Prof. Salameh was awarded the First Class Order of Merits from the president of the Federal Republic of Germany in 2006 and has been the recipient of many other local and international orders of merit. His main research interests are hydrogeology, hydrochemistry, applied geology, and geophysics.

Dr. Musa Shteiwi is a professor of sociology and currently the director of the Center for Strategic Studies at the University of Jordan. He obtained his Ph.D. from the University of Cincinnati, Ohio, in the United States in 2011. He has more than 25 years of experience in teaching and research in the areas of political sociology, human rights, development, and gender. He has also served as an advisory consultant for the government, the UN, the World Bank, and other international organizations and research institutions and has provided technical support on social policies for Egypt, Bahrain, Kuwait, Oman, and Jordan. He received the State Encouragement Award for his research on gender. He has written over 35 papers and published books on development, social policies, poverty, unemployment, women, social classes, civil society, political parties, and youth.

Dr. Marwan Al-Raggad is associate researcher at the Water, Energy and Environment Center at the University of Jordan. He holds a Ph.D. in Groundwater Management and Post Doc in Groundwater Modelling and has solid experience in water management gained from his work as a hydrogeologist at the Ministry of Water and Irrigation from 2002 to 2010. Since joining the University of Jordan in 2010, Dr. Al-Raggad has led many international research projects in the domain of climatic change effects on water resources, managed aquifer recharge, groundwater quality, and treated wastewater reuse in ground water recharge.

List of Abbreviations

ACSAD	Arab Centre for the Studies of Arid Zones and Dry Lands
BOD_5	Biochemical oxygen demand over five days
BGR	Bundesanstalt für Geowissenschaften und Rohstoffe
°C	Degrees Celsius
Ca^{2+}	Calcium
CBZ	Carbamazepine
Cl^-	Chloride
cm	Centimeter
COD	Chemical oxygen demand
d	Day
dS/m	Decisiemens per meter
EC	Electrical conductivity of water
E. coli	*Escherichia coli*
g	Gram
GEF	Global Environment Facility
ha	Hectare
HCO_3^-	Bicarbonate
JVA	Jordan Valley Authority
JD	Jordanian Dinar
K^+	Potassium
KAC	King Abdullah Canal
Km	kilometer
KTD	King Talal Dam
L	Liter
L/c.d	Liter per capita and day
m	Meter
masl	Meter above sea level
mbsl	Meter below sea level
MCM	Million cubic meters
meq/L	Milliequivalents per liter
Mg^{2+}	Magnesium

mg/L	Milligram per liter
mSv	Millisievert
mm	Millimeters
mmhos/cm	Millimhos per centimeter
μS/cm	Micro Siemens per cm
WERSC	Water and Environmental Research and Study Centre
WSP	Waste stabilization ponds
WWT	Wastewater treatment
WWTP	Wastewater treatment plant
MWI	Jordan Ministry of Water and Irrigation
Na^+	Sodium
nCi/L	Nanocurie per liter
$NH_4^+\text{-}N$	Ammonium-nitrogen
NGOs	Non-governmental organizations
$NO_3^-\text{-}N$	Nitrate-nitrogen
$NO_2^-\text{-}N$	Nitrite nitrogen
NRA	Jordanian Natural Resources Authority
PC	Pharmaceutical compounds
PEC	Pollutant emerging concern
pH	Hydrogen ion activity
PO_4^{3-}	Phosphate
ppm	Parts per million
ppt	Precipitation
SO_4^{2-}	Sulfate
TN	Total nitrogen
TDS	Total dissolved solids
TSS	Total suspended solids
UNEP	United Nations Environment Program
WAJ	Water Authority of Jordan
WHO	World Health Organization
yr	Year

Chapter 1
Introduction

As a naturally semi-arid country, Jordan has limited amounts of rainfall and hence limited surface and groundwater resources.

The water shortage is perceived as a straightforward population-induced scarcity of resources aggravated by quality deterioration and resources misallocation, processes which in themselves negatively reflect on the availability of the naturally scarce resources.

Population growth, industrialization, irrigation projects and improving standards of living over the last few decades have not only led to increasing water use and over-exploitation, but also to deteriorating water qualities as a result of the various human activities.

This situation has prompted a number of research projects and studies, conferences, workshops etc. at many levels and institutions, notably at the University of Jordan and the Ministry of Water and Irrigation. Such activities have been highly appreciated and well received by universities, scientific institutions, research centers, and national, regional and international organizations both in Jordan and abroad.

Through the resulting publications, new information, analyses, facts and methodologies have been made available to all concerned. The way of looking at the water sector has changed since the beginning of the twenty-first century. The water sector now requires advanced socio-economic, strategic and environmental approaches because it has surpassed the stage of allocating more resources to cover the demand. Considering water as an issue of national strategic significance has become an imperative for Jordan: Hence the relevance of this book, which looks at the water sector in an integral way taking into consideration all socio-economic, political and strategic options.

© Springer International Publishing AG, part of Springer Nature 2018
E. Salameh et al., *Water Resources of Jordan*, World Water Resources 1,
https://doi.org/10.1007/978-3-319-77748-1_1

1.1   Country Profile

Information below is based on data obtained from the Department of Statistics (DOS open files 2017).

Area: 89.400 km², (Fig. 1.1).
Population (2016): Jordanian Nationality 6.5 million, guest workers 850 thousand, Syrian refugees 1.2 million, other refugees 150 thousand.
Rate of natural growth of Jordanian nationals: 2.4% per year.
Economic sectors: Agriculture ≈ 10%, industry 22%, services 68%.
Labor force: Agriculture ≈ 11%, industry 27%, services 62%.
Literacy rate: ≈ **88%**.

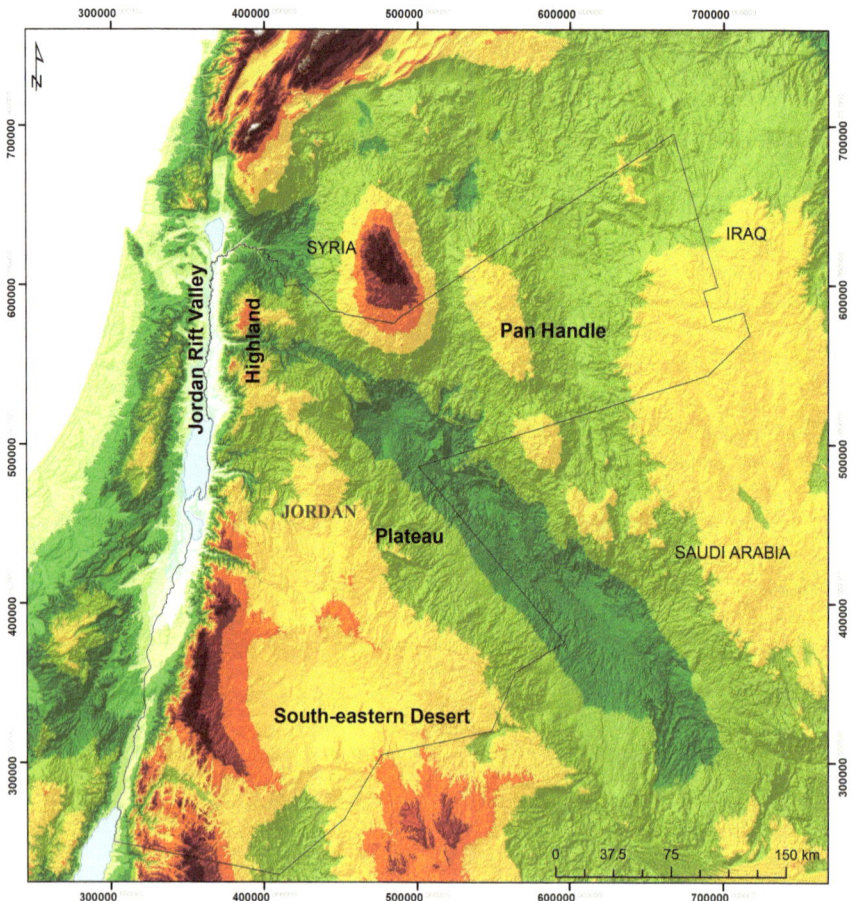

Fig. 1.1 Location map of Jordan showing the Jordan Rift Valley, Wadi Araba, highland, plateau pan handle

Exports: Potash, phosphate, fertilizers, small and intermediate industrial products, medicine, manpower, vegetables and fruits, services such as medical care and expertise.

Imports: Fuel, food (grain, meat, etc…), vehicles, heavy machinery, industrial plants, wood, iron, paper … etc.

Energy: Only very limited gas fields, large oil shale deposits which are not yet mined.

Food production: covers around 50% of the country's needs.

1.2 Topography

Based on Salameh and Banayan (1993) and Salameh (1996), the country consists of different distinctive topographic units trending in a general north-south direction. The major geologic event which incorporated rifting along the Jordan Rift Valley line during the last 20 million years caused the formation of the Rift Valley with the highlands on both sides; east and west are responsible for the present topographic configuration of the country. The eastern highlands in Jordan slope to the steppe in the east.

The Rift Valley trends in a general south-north direction and extends from the Gulf of Aqaba at sea level to around 240 masl at a distance of 80 km to the north, to the Dead Sea at 430 mbsl and then to Lake Tiberias at 210 mbsl. The bottom of the Dead Sea lies at around 750 mbsl (Neev and Emery 1967).

The Rift Valley with a length of 375 km has a width of about 30 km in the area of Aqaba and Wadi Araba and narrows to around 15 km in the Dead Sea area and to 4 km south of Lake Tiberias.

The eastern shoulder of the Jordan Rift Valley rises to more than 1000 masl in the north in Ajlun and Balqa mountains; and to more than 1200 masl in Shoubak and Ras El Naqab areas with a width ranging from 30 to 50 km and extending from the Yarmouk River in the north to Aqaba in the south. These highlands slope gradually to the plateau in the east at elevations of 600–800 masl with the deepest part of this plateau at an elevation of 500 masl in Azraq area, and slopes more sharply towards the Rift Valley in the west. The mountains forming the highlands consist mainly of sedimentary rocks with deeply incised wadis draining towards the Rift Valley in a westerly direction.

The plateau has hills and weakly incised wadis, but in general it possesses smooth topography. Surface runoff water, which does not flow to the Jordan Rift Valley, discharges into desert playas or Qa'as forming extended shallow lakes in winter time and dry mud flats in summer time.

In the north-eastern part of the country a flat plateau, the Panhandle, with very smooth topography rises from 500 m in Azraq area to about 900 m at the Jordan-Iraq borders. Its western parts are formed by Jabal Arab-Druz (Horan) volcanic mountains which rise to about 150 m above the plateau level.

The southern desert forms also a flat area where the topography rises in its south-western parts to more than 1500 masl (Aqaba Mountains).

The most south-eastern part of the plateau, south-east of Ras El Naqab escarp-ment, is considered a different topographic unit and although it belongs to the same plateau it is separated from the plateau by Ras en Naqab escarpment. The elevation of the area is around 900 masl, with a north-south width of around 100 km forming what is referred to as the Southern Desert of Jordan. It is strongly dissected by deep wadis in the western part and smooth wadi slopes in the eastern part.

1.3   Climate

The text on climatic settings is based on Salameh and Banayan (1993) and Salameh (1996) with modified maps and figures obtained from the Department of Meteorology (DOM 2016).

Jordan lies in the semi-arid area of the world with the exception of the highlands, with a width of around 30 km and a length of some 300 km which enjoy a Mediterranean type climate.

Temperatures in the Jordan Rift Valley can rise in summer to 45 °C with an annual average of 24 °C. In winter the temperature in this area reaches a few degrees above zero. Frost is a rare event, but it occurs from time to time.

The highlands enjoy a temperate climate with cold and wet winters with tem-peratures reaching a few degrees below zero during the night, and hot and dry sum-mers with temperatures reaching 35 °C at noon and with a relative humidity of 15–30%. During the summer, temperatures at night normally drop to less than 20 °C accompanied by the formation of dew.

The eastern and southern areas are hot in summer and cold in winter with tem-perature during summer days of more than 40 °C dropping in winter to a few degrees below zero, especially during the night. The relative humidity is generally low; in winter it reaches 50–60%, and in summer it sometimes drops to 15%.

Throughout most of the year the relatively low humidity makes the hot summer days more tolerable and the cold winter days more severe.

1.4   Precipitation

Precipitation in Jordan normally occurs in the form of rainfall with snowfall occur-ring generally once or twice a year mostly over the highlands. The rainy season extends from October to April, with the highest precipitation amount falling during January and February. Precipitation becomes less pronounced the less rainfall an area receives (Fig. 1.2).

The highest precipitation amounts fall over the highlands of Ajlun, Balqa, Karak and Shoubak which receive long-term annual averages of 600, 550, 350 and

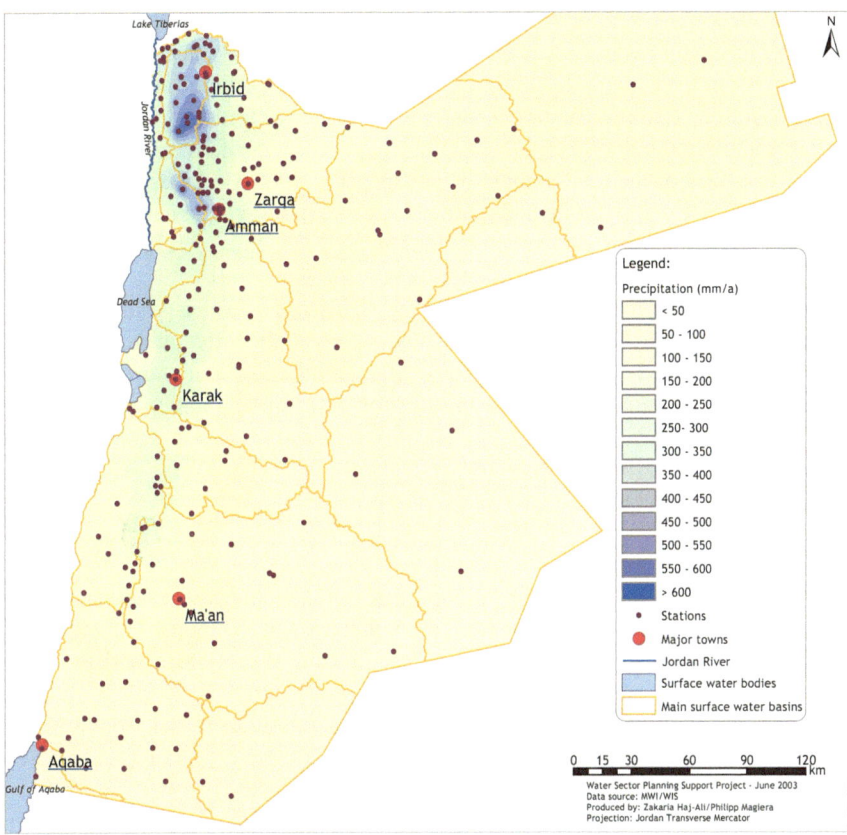

Fig. 1.2 Locations of weather stations in Jordan with the ranges of (30–70 years) average amounts of precipitation

300 mm. Precipitation decreases drastically to the east of the highlands, and more strongly to the west (Fig. 1.2). For example it decreases from an average of 600 mm/year in Ajlun to 250 mm/year in the Jordan Valley to the west within a distance of 10 km and a difference in altitude of 1200 m. The decrease in the easterly direction is less than due west; for example, from 300 mm/year in Shoubak to 50 mm/year some 30 km to the east in Jafr area.

Generally, the following facts can be stated about precipitation in Jordan:

– Jordan's territories receive an average annual amount of precipitation of 7200 MCM increasing to 12,000 MCM in a wet year and decreasing to 6000 MCM in a dry year.
– An average annual precipitation of more than 500 mm is received by around 1.3% of Jordan's area, between 300 and 500 mm is received by 1.8%, between 200 and 300 mm by 3.8%, between 100 and 200 mm by 12.5% and the rest of the area receives less than 100 mm/year.

A clearer and more accurate picture of Jordan's water situation is perceived when knowing that only about 3% of the total area of the country receives an average amount of ppt exceeding 300 mm/year. This is the least amount needed to grow wheat under dry farming conditions under the prevailing climate of the country.

Since a minimum of 300 mm/year rain is required for dry farming it can be concluded that 83% of the total amount of precipitation falls over areas which cannot be used in rain-fed agriculture and that only 17% of precipitation can be useful for that purpose. The other part of precipitation, 83% requires expensive technical interventions to make it partly available.

Part of the precipitation water flows along wadis and is collected in dams or in desert playas and part of it percolates down to replenish the groundwater resources.

Being a semi-arid country, atmospheric dust and the low amounts of precipitation are generally reflected in increasing salt contents of precipitation water.

1.5   Evaporation

The prevailing semi-arid conditions in Jordan govern not only precipitation amounts but also the potential evaporation (Fig. 1.3), which rises from about 1600 mm/year. in the north-western highlands of the country to more than 4000 mm/year. in the south-eastern desert areas.

Along the Jordan Rift Valley the potential evaporation decreases from a maximum of 4000 mm/year. in the Aqaba region to some 2500 mm/year. in the Dead Sea area and to 2000 mm/year. in the north to the south of Lake Tiberias. These potential evaporation rates are 5–80 times the average amounts of precipitation over these areas.

Potential evaporation from the plateau areas increases in easterly and southerly directions: from an average of 3000 mm/year. at the eastern foot of the highlands to around 4000 in the center of the plateau. The rates in the south-eastern deserts are 3500–4400 mm/year.

The potential evaporation in the plateau area and in the south-eastern desert areas are 12–100 times the amount of precipitation received in these areas.

The high evaporation potential in Jordan makes precipitation, especially in the eastern and southern parts of the country, ineffective because precipitation water evaporates immediately after precipitation leaving soils deprived of their moisture content and hence, it does not allow the development of plants and green lands.

High evaporation rates, low precipitation amounts and relatively high salt contents in precipitation water lead generally to salt concentrations in flood and recharge water.

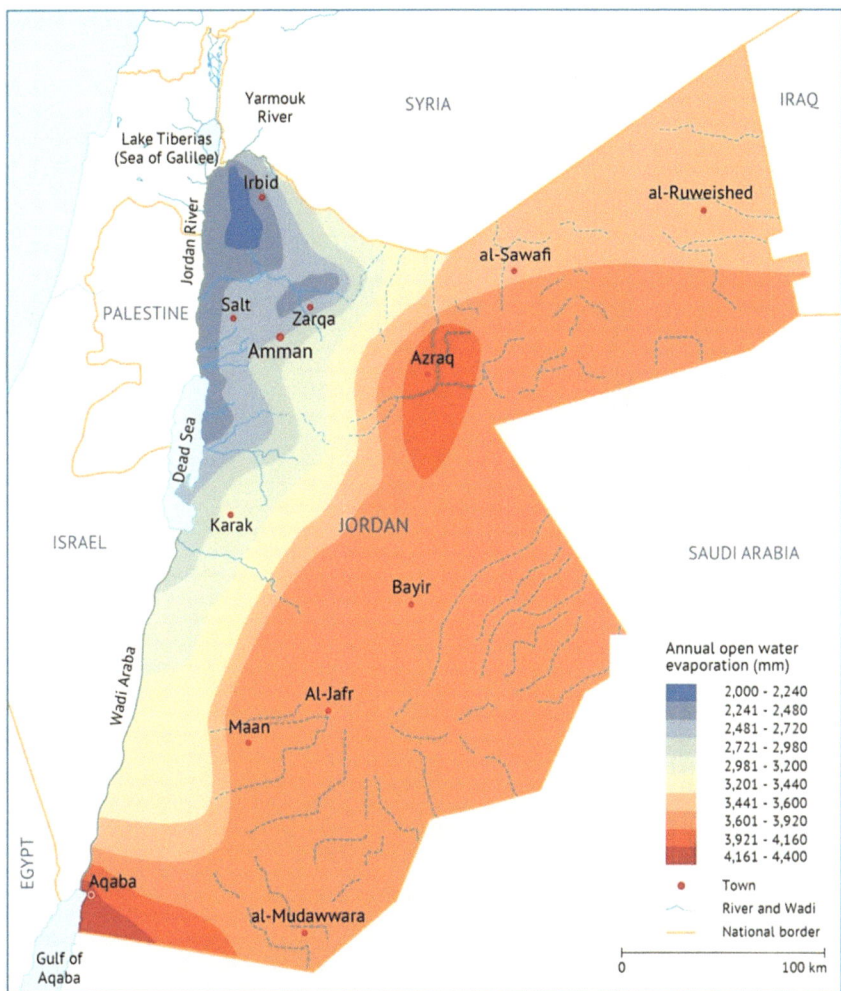

Fig. 1.3 Potential annual evaporation in Jordan

References

DOM (Department of Meteorology) (2016) Publications and files of the Department of Meteorology/Amman

Neev D, Emery DO (1967) The dead sea depositional processes and environments of evaporates, Ministry of Development, Geol. Survey of Israel, 41, 147p, Jerusalem

Salameh E (1996) Water quality degradation in Jordan. Friedrich Ebert Stiftung, Amman and Royal Society for the Conservation of Nature, Amman, 179 p

Salameh E, Bannayan H (1993) Water resources of Jordan – present status and future potentials. Friedrich Ebert Stiftung, Amman, 183 p

Chapter 2
Water Resources

The development of the different social and economic sectors during the last six decades has been accompanied by increasing water extraction and use. Therefore two types of water situations have to be differentiated:

- Pre-development water situation and
- Actual water situation

The main reasons for this differentiation are:

1. Excessive groundwater extractions which have strongly affected river and wadi base flows.
2. Treatment of waste waters and the discharge of their effluents along wadis to collect in dams.
3. Climatic changes which are thought to have negatively impacted surface and groundwater resources.

This chapter discusses the pre-development water sector situation and describes the actual water situation.

2.1 Surface Water Resources

Jordan has only one major river the Jordan River which in the 1940s and 1950s used to discharge around 1400 MCM/year into the Dead Sea. This river is a very small one compared with international rivers like the Nile or Euphrates, because its total annual discharge amounts to only about 1.5% that of the Nile and 4.3% that of the Euphrates.

The text of this chapter is based on Salameh and Banayan (1993) and Salameh (1996) with modified maps and figure (MWI 2016).

© Springer International Publishing AG, part of Springer Nature 2018
E. Salameh et al., *Water Resources of Jordan*, World Water Resources 1,
https://doi.org/10.1007/978-3-319-77748-1_2

Surface water resources are found in the Yarmouk and Zarqa rivers and in wadis like Karak, Mujib, Hasa, Yabis and El-Arab, in addition to flood flow in wadis in the different parts of the country. Figure 2.1 shows the surface water basins in Jordan and Fig. 2.2 shows the catchment area of the Jordan River which extends into Jordan, Syria, Lebanon, Palestine and Israel making the availability of its water dependent on upstream countries use in their parts of the catchment.

2.1.1   The Jordan River Area

The Jordan River
The catchment area of the Jordan River measures 18.194 km^2 with about 2833 km^2 lying upstream of Lake Tiberias outlet. The eastern catchment area, downstream of Tiberias, measures 13.027 km^2, and the western 2344 km^2.

Three main springs feed the headwaters of the Jordan River; namely, Hasbani in Lebanon, Dan in Israel, and Banias in Syrian territory occupied by Israel. The three streams flow together to form the Upper Jordan River. The surface catchments of the springs are relatively small considering the large quantities of water discharged from them; therefore, it is assumed that their underground catchments extend further to the north beyond the surface catchment, into Syria and Lebanon.

The discharge of the Jordan into the Dead Sea – prior to the development of its water resources in Jordan, Syria and Israel – was 1370 MCM/year. At the present time this amount is not more than 150–200 MCM/year mostly consisting of irrigation return flows, undammed inter-catchments or saline spring discharges.

The saline springs in the immediate surroundings of Lake Tiberias and at its bottom, discharging around 16 MCM/year with a salt content of around 6000 mg/L, are channelled downstream of Lake Tiberias into the headwaters of the Lower Jordan River.

In the 1950s and 1960s, prior to the use of its water by the different riparian states, the Yarmouk River used to discharge around 500 MCM/year into the Jordan River. Over the last three decades, this amount has gradually declined to discharges as a result of large floods which cannot be accommodated by the existing extraction facilities in Syria. Between 2007 – the year of its construction on the river – and 2013, the Wahda (Unity) dam collected only 10–20 MCM/year although the design capacity is 110 MCM. This is a direct result of the Syrian extractions from the headwaters of the river, although the historic flow of the river at the Unity Dam site (Maqarin) used to average 260 MCM/year. After 2013, due to some damage to Syrian water facilities within the Yarmouk catchment the discharge of the river increased to 30–50 MCM/year.

The riparian countries of the Jordan River have, over the last few decades, diverted other wadis and springs of the Jordan Valley. The present flow of the Jordan

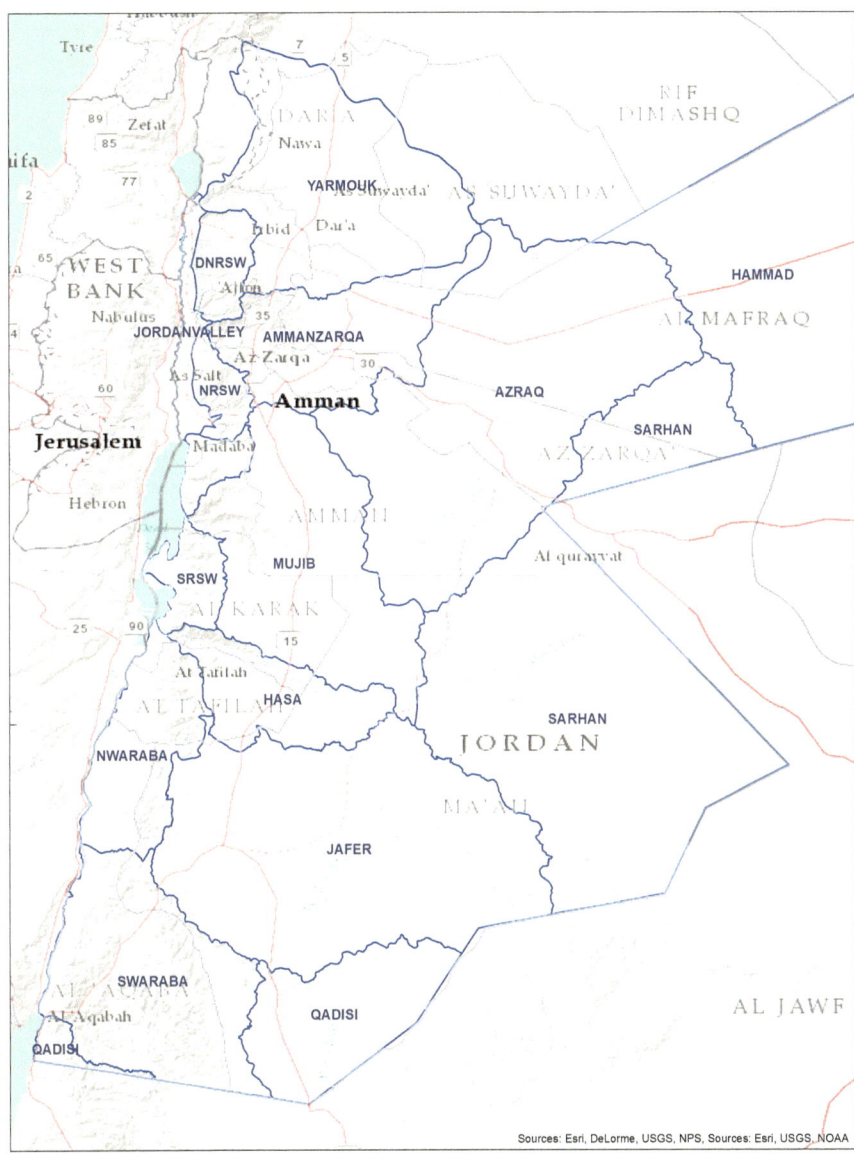

Fig. 2.1 Main surface water basins in Jordan (Hammad, Azraq, Jafr, Sirhan and Disi form closed basins, Wadi Araba South drains into the Red Sea; Wadi Araba North, Hasa, Mujib, Karak, Zarqa Main, Southern Rift, Northern Rift, Jordan Valley, Yarmouk and Zarqa Rivers drain into the Dead Sea)

Fig. 2.2 Catchment area of the Jordan River

River consists of runoffs due to rain over areas downstream of water collection constructions, irrigation return flows or saltwater springs.

Yarmouk River
The Yarmouk River flows along the borders between Jordan and Syria and joins the Jordan River in a border area with Israel. The total catchment area of the river measures 6790 km^2; 1160 km^2 within Jordan and 5630 km^2 in Syria (Fig. 2.3).

Along its course from the foothills of Hermon and Jabel Arab-Druz areas to its confluence with the Jordan River, different wadis and creaks feed the Yarmouk River. Tributaries of Harir, Allan and Raqqa in Syria and Shallala and El Humra in Jordan are the most important contributors to the river flow in terms of water quantities.

The catchment area of the Yarmouk River is agrarian, containing small types of industries in both Jordan and Syria. Effluents of some waste water treatment plants reach the river during floods. Also leachates of some solid waste disposal sites directly reach the river course during rain events when their liquid loads exceed evaporation and infiltration rates.

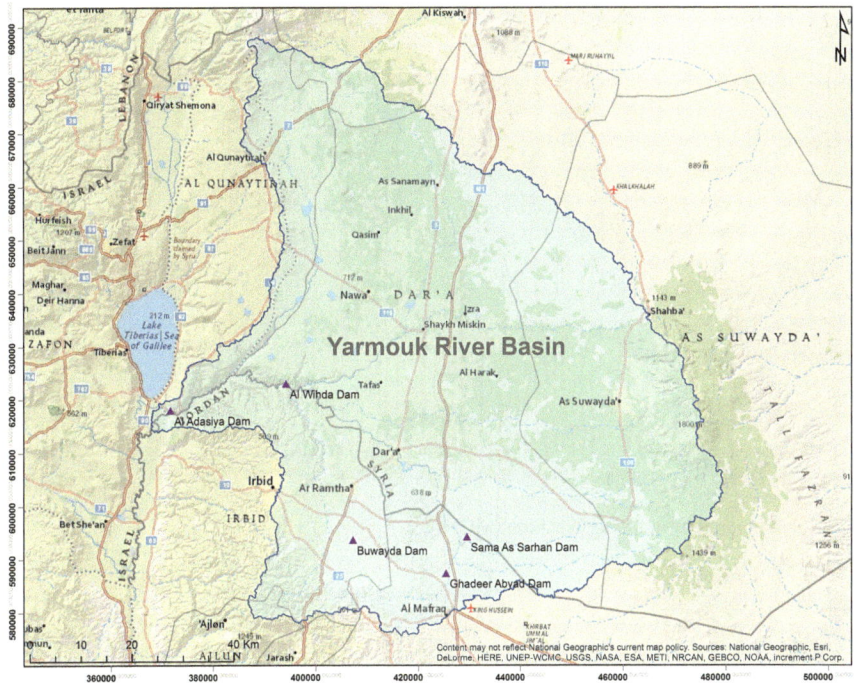

Fig. 2.3 Catchment area of Yarmouk River in Jordan and Syria

The average annual rainfall over the catchment area is 372 mm/year. The north-western parts of the catchment bordering the Hermon Mountains receive precipitation amounts of more than 1000 mm/year decreasing to 250 mm/year in the south-eastern area of the catchment.

During the pre-development era of the 1940s, 1950s and 1960s the Yarmouk River at Adasiya used to discharge an average of 467 MCM/year (1927–1964). More recent measurements, although masked by unknown usage by the riparian states, show a drastic decline in the river discharge as a result of increasing extractions of groundwater which have lead to declining base flow (Figs. 2.4, 2.5 and 2.6). Decreasing precipitation over the last five decades has also contributed to the decreasing discharges.

In the period 1950–1976 the river discharged an average of 400 MCM/year. Recent estimates of the catchment water resources indicate an average total amount of natural water resources of around 360 MCM/year. Most of it is extracted by Syria and Israel before reaching the Yarmouk River.

Rain-fed, some irrigated agriculture and the sparse population of the catchment with limited small industries is reflected in the water quality of the Yarmouk River and pollution parameters can only be measured during low river flows.

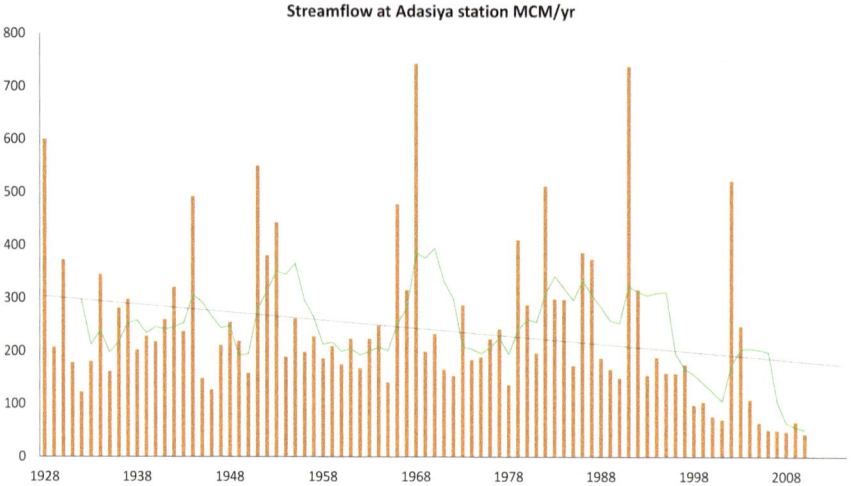

Fig. 2.4 Stream flow at Maqarin station (MCM/year)

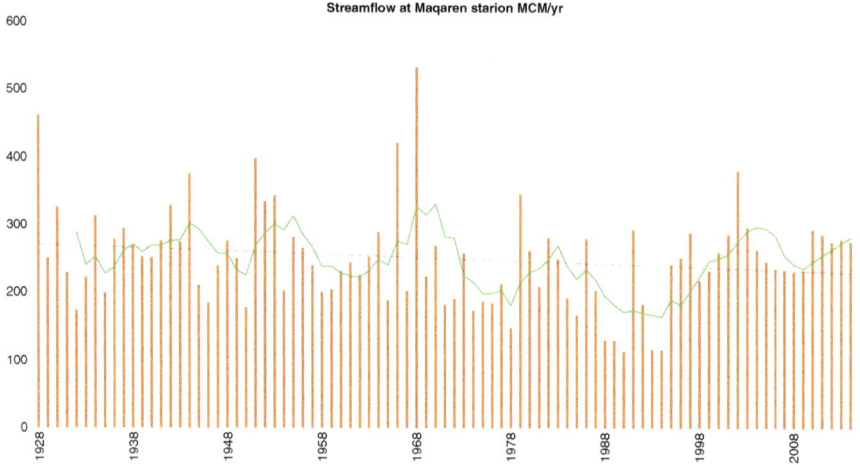

Fig. 2.5 Stream flow at Adasiya station (MCM/year)

Zarqa River

The catchment area of Zarqa River measures 4025 km^2 and extends from the foothills of Jabel Druz in Syria to the Jordan River (Fig. 2.7). It is the second largest river in Jordan in terms of its catchment area and its mean annual discharge.

The river is formed by two main branches; Wadi Dhuleil, which drains the eastern part of the catchment area, and Wadi Zarqa, which drains the western part. Both meet at Sukhna to form the Zarqa River. Wadi Dhuleil used to drain only flood flows as a result of precipitation and Wadi Zarqa used to drain flood and base flows.

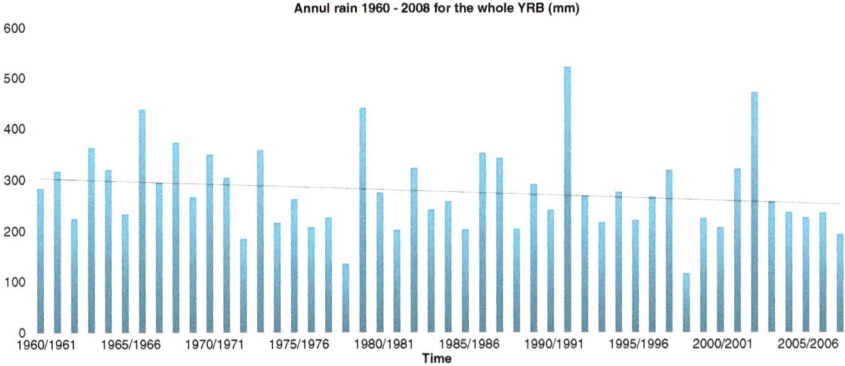

Fig. 2.6 Rainfall long-term record for the Jordanian part of the Yarmouk River basin

Fig. 2.7 Catchment area of Zarqa River

Zarqa River catchment area is the most densely populated area in Jordan, comprising around 65% of the country's population and more than 65% of its industries.

Household and industrial waste waters are generally sewered and treated in different waste water treatment plants before discharge into the surface water system of Zarqa River.

Solid wastes produced within the catchment are dumped in disposal sites located in the catchment area. Their leachates end up in surface and groundwater resources, causing local pollution and threatening to contaminate the aquifers.

The average annual amount of precipitation over the catchment area reaches 237 mm. The eastern part, around half of the total catchment area, receives an average amount of 182 mm/year. The middle part receives an average of 243 mm/year and the western part, comprising the highlands and the slopes to the Jordan Valley area, receives an average of 397 mm/year.

Snow generally falls once or twice a year over the highlands, and the eastern part of the catchment receives almost only rainfall.

The highest amounts of precipitation are received in the highlands of Salt and Amman with an average of 550 mm; increasing in a wet year to 750 mm and decreasing in a dry year to 350 mm. In the semi-arid most eastern part of the catchment precipitation averages, in a normal year, 80 mm, increasing to 150 mm in a wet year and decreasing to 50 mm in a dry year.

The potential evaporation in the highlands of Salt and Amman is about 1600 mm/year increasing to about 2300 mm/year in the eastern part of the catchment. The conclusion that can be drawn from the distribution of precipitation and potential evaporation is that there is not enough water to satisfy the needs of the evaporation force of the climate, which is far less pronounced during the winter months than during the summer months, thus, allowing precipitation water to infiltrate and recharge the groundwater during the rainy season.

In the years 1950–1976 the Zarqa River at Deir Alla used to discharge an average of 64.88 MCM/year. After 1976, the natural system of the river was changed by different factors such as the construction of the King Talal Dam on the Zarqa River, discharges of treated and untreated waste water into the water courses and imports of water for domestic and industrial uses into the catchment area.

Such activities caused major increases in the river flow on the one hand and negatively affected its water quality on the other. Due to incidental groundwater recharge with treated waste water, the river base flow has doubled during the last 30 years (Fig. 2.8).

King Talal Dam on the Zarqa River was constructed in 1977 to accommodate 56 MCM of flood and base flows and this was raised to 89 MCM in 1988. The natural flood and base flows of the Zarqa River are not enough to fill the dam in an average year. But since increasing amounts of water were imported into the catchment area to cover the increasing demand, the treated waste water effluents discharged within the river catchment now reach the dam and fill it almost every year.

At present treated waste water contributes around 70% of the river discharges.

During the pre-development era the natural water quality of Zarqa River was good enough for household uses, but since then it has been negatively affected by pollutants so that, at present, it can only be used for irrigating certain crops. The water quality of the river during the rainy season is still acceptable for most uses, but in summer times with no flood water the quality degrades and the water can only be used in restricted irrigation.

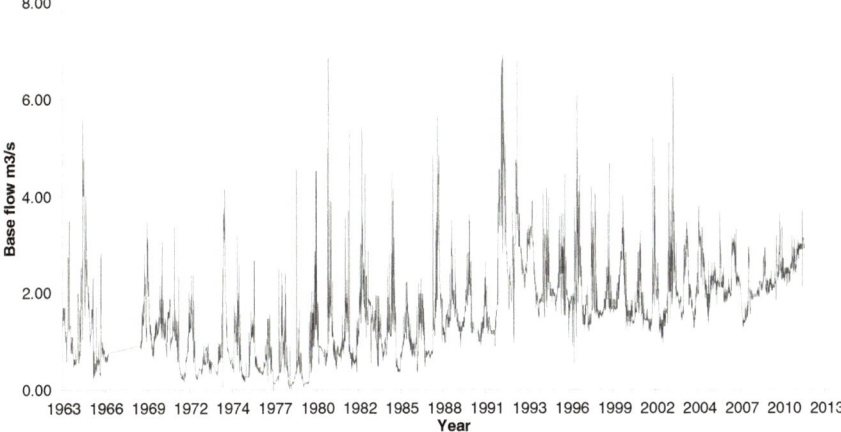

Fig. 2.8 Zarqa River base flow as recorded at Jarash bridge station AL0060

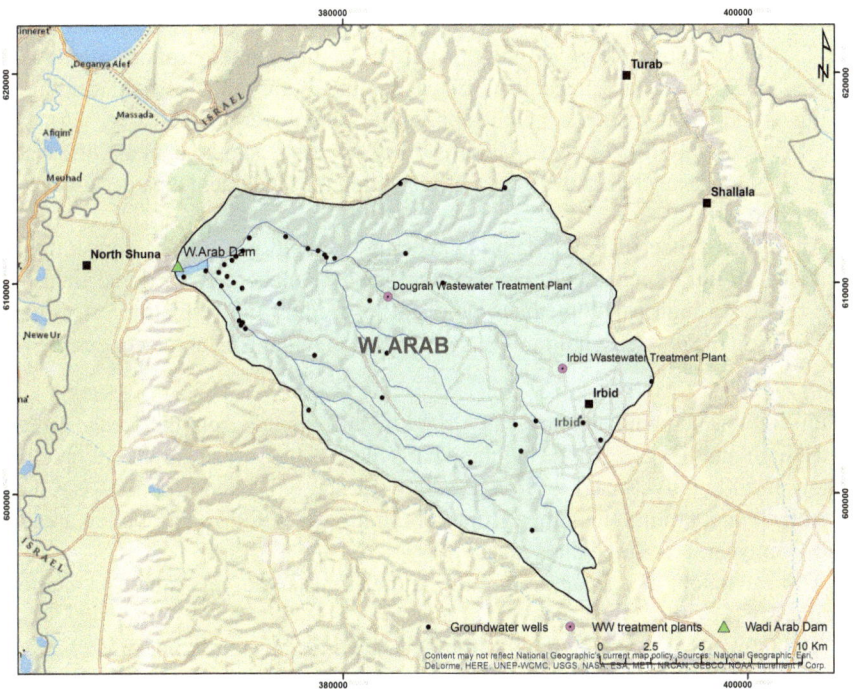

Fig. 2.9 Catchment area of Wadi Arab

Wadi El-Arab

Wadi El-Arab catchment borders the Yarmouk catchment and measures 267 km^2 (Fig. 2.9). The average amount of precipitation over the highlands of the catchment averages 500 mm/year and decreases to 350 mm in the Jordan Valley area. The potential evaporation ranges from 2000 mm/year along the highlands to 2400 mm/year in the Jordan Valley area.

The pre-developmental average water discharge of the wadi was around 31.4 MCM/year, with around 25 MCM/year consisting of base flows.

Wadi El-Arab Dam was built in 1987, with a total capacity of 20 MCM to collect flood and base flows for use in irrigation in the Jordan Valley area. Waters originating within its wadi catchment filled the dam only in the very few wet years such as 1991/92. In other years, the dam served as a storage reservoir for water pumped from King Abdullah Canal in the Jordan Valley during floods.

The city of Irbid expanding westward into the catchment area may put increasing pressure on the quality of the water collected in the dam.

Two waste water treatment plants for Irbid City were constructed in the upper reaches of Wadi El-Arab. And although the effluents of the treatment plants were piped to bypass the dam, floodwaters still enter the treatment plant and wash its untreated contents and the wastes along Wadi El-Arab into the dam reservoir, negatively affecting its water quality.

Groundwater extraction upstream of the dam has, during the last three decades, resulted in groundwater level declines and hence the cessation of groundwater natural discharges from springs. During this period, the drop in the groundwater levels exceeded 120 m – a fact that calls into question the future reliability and durability of this drinking water source which supplies the Irbid governorate.

At present only flood flows of about 5.5 MCM/year reach the dam and the wadi does not receive any base flow from springs within the catchment area because they have dried out.

The water collected in the dam is generally of good quality. The conventional treatment of filtration and chlorination is sufficient to make it fit for domestic uses. The relatively high trihalomethane potential, especially the formation of bromoform during the dry season upon water chlorination remains of some concern.

Wadi Ziglab

Wadi Ziglab catchment area measures 106 km² and extends from the Jordan Valley to the highlands (Fig. 2.10). Its highland parts receive an average amount of precipitation of 500 mm/year and its Jordan Valley parts only around 300 mm/year. The potential evaporation rates range from 2050 mm/year in the Jordan Valley area to 2200 mm/year in the highlands area.

The catchment area is agrarian with natural forests and very small population numbers.

The total discharges of springs within the catchment average some 5 MCM/year and flood flows also average some 5 MCM/year.

Wadi Ziglab Dam with a total capacity of 4.3 MCM was constructed in 1966 with the aim of using its water for irrigation in the Jordan Valley area.

The hydrologic situation in the catchment area has not changed much over the decades. The area has been affected by increased urbanization and climatic changes, but these are not yet strongly reflected in the flow regime of this wadi.

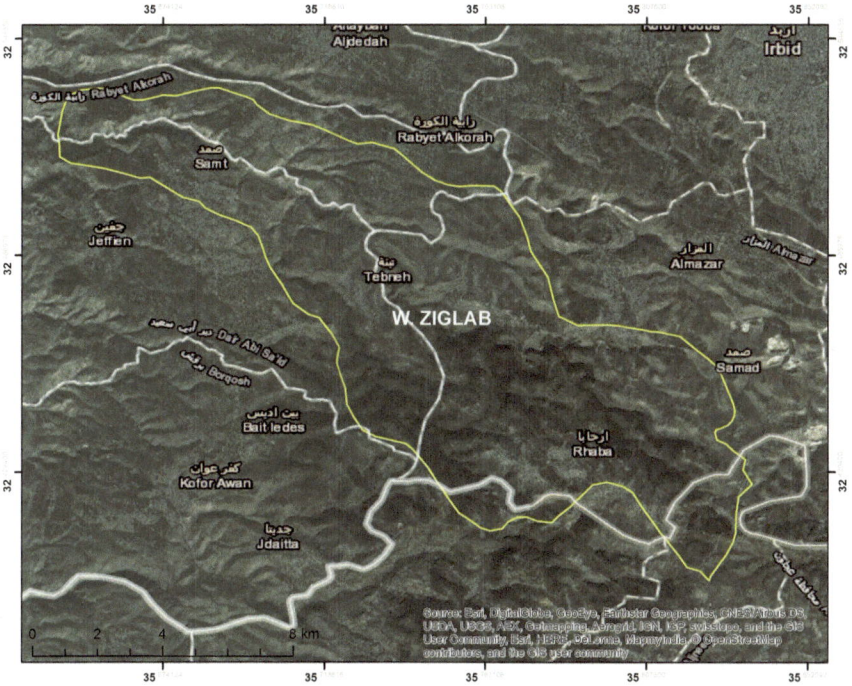

Fig. 2.10 Catchment area of Wadi Ziglab

Wadi Shueib

The catchment area of Wadi Shueib measures approximately 180 km² extending from Suweileh in the west at elevations of about 1200 m down to the Jordan Valley at 380 mbsl (Fig. 2.11). Precipitation over the catchment area ranges on average from 500 mm/year in the high mountains to 150 mm in the Jordan Valley area. Along the highlands it partly falls in the form of snow. The potential evaporation ranges from 2700 mm/year in the highlands to 2500 mm/year in the Jordan Valley area.

The average natural flow of the wadi is 5.7 MCM/year consisting of 1.8 MCM/year flood flows and 3.9 MCM/year base flows. In addition, the effluents of the waste water treatment plants of Salt, Fuheis and Mahis towns discharge to the wadi and collect in the dam at the entrance of the wadi into the Jordan Valley. This dam was constructed in 1968 with a capacity of 2.3 MCM with the aim of using its water for irrigation in the Jordan Valley area.

The pumping of spring water and extractions from the groundwater have led to decreasing base flows of the wadi, but spring water which is not pumped continues to flow along the wadi and is used in irrigation.

Wadi Kafrain

The catchment area of Wadi Kafrain measures 189 km² lying at elevations ranging from 1200 masl in the highlands down to areas lying 350 mbsl in the Jordan Valley

Fig. 2.11 Catchment area of Wadi Shueib

(Fig. 2.12). Precipitation along the highland parts of the catchment averages 550 mm/year and may fall in the form of snow, whereas in the Jordan Valley parts the average precipitation reaches only 140 mm/year and falls entirely in the form of rain. The potential evaporation rates range from 2700 mm/year in the highland parts to 2400 mm/year in the Jordan Valley parts.

During the pre-developmental era the average natural discharge of Wadi Kafrain was 6.4 MCM/year, consisting of 1.6 MCM/year flood flow and 4.8 MCM/year base flow. Effluents of Wadi Sir waste water treatment plant are discharged into Wadi Kafrain or its tributary wadis.

Treated and untreated waste water from different towns and villages, like Wadi Sir and Na'ur is also discharged along Wadi Kafrain or its tributaries.

In 1968 a dam with a capacity of 3.8 MCM was constructed at the entrance of Wadi Kafrain into the Jordan Valley to provide water for irrigation. The dam was raised to a capacity of 7.5 MCM in 2007.

At present, the dam collects flood flows, irrigation return flows, treated and untreated waste waters and groundwater discharged from artesian wells drilled into the lower pressurized aquifer in the upstream area of the dam. It receives good quality water from artesian wells, medium quality water from floods mixed with treated and untreated waste waters and bad quality water from irrigation return flows. The dam's water is used only for irrigation.

Fig. 2.12 Catchment area of Wadi Kafrain

Wadi Kufranja

The catchment area of Wadi Kufranja measures 111 km^2 and extends from Ajlun Mountains at about 1300 masl down to the Jordan Valley at 225 mbsl. Precipitation over Ajlun averages 600 mm/year partially recieved in the form of snowfall; whereas in the Jordan Valley area precipitation averages 300 mm/year and falls as rain. The base and flood flow amounts in an average year measure 6.82 MCM, of which around 5.8 MCM/year are base flows.

A dam was constructed on the wadi to collect flood and base flows, but it now also receives the effluents of Kufranja waste water treatment plant and receives less base flow than expected because of the pumping of the springs' water for municipal uses.

Other Wadis Discharging into the Jordan Valley

These wadis are not dammed and include Yabis, Jurum, Rajib, Hisban and other small catchments.

The rainfall in these areas ranges from 150 mm/year in the Jordan Valley area up to 550 mm/year over the highlands, with potential evaporation rates ranging from 2100 mm/year in the highlands to 2700 mm/year in the Jordan Valley area.

Most of the base flows of these wadis are used in irrigation both within their courses and partly at the foothills of the Jordan Valley. Their flood flows still reach the Lower Jordan River. Plans are underway to construct suitable water harvesting structures on these wadis.

2.1.2   Dead Sea Wadis

Wadi Zarqa Ma'in

The catchment area of Wadi Zarqa Ma'in measures 272 km² and extends from the highlands at an elevation of about 1000 masl to the level of the Dead Sea at 430 mbsl (2017). Precipitation over the catchment falls in the form of rain and ranges from 350 mm/year in the highland surroundings to 100 mm/year at the shores of the Dead Sea. The potential evaporation rates range from 2900 mm/year in the highlands to 2400 mm/year at the shores of the Dead Sea.

The discharge of Wadi Zarqa Ma'in into the Dead Sea averages 23 MCM/year, of which only around 3 MCM/year flows as floods and 20 MCM/year as base flow. Thermal water issuing from a number of springs ranging in discharge from seepage size to 150 L/s constitutes a major part of the base flow with salinities of around 3000 µS/cm. The wadi water is collected at its entrance into the shore area of the Dead Sea and is diverted for the municipal water supply in the Amman area.

Wadi Mujib (Including Hidan)

Wadi Mujib catchment area measures 6.596 km² and ranges in elevation from 1100 masl in the highlands to 430 mbsl at the shores of the Dead Sea (Fig. 2.13). Precipitation over the catchment area falls in the form of rain and seldom in the form of snow over the highlands and ranges from 350 mm/year along the mountainous highlands to 100 mm/year at the shores of the Dead Sea. The potential evaporation ranges from 2450 mm/year at the shores of the Dead Sea to 3500 mm/year in the highlands.

In the downstream area of the confluence of Wadi Hidan with Wadi Mujib the discharge averages 83 MCM/year directly discharging into the Dead Sea, half of which consists of base flow and the other half of flood flows.

In the lower reaches of the wadi the average base flow of around 30 MCM/year mostly consists of lightly mineralized water issuing from the sandstone aquifer complex. The salinity of some springs reaches 2000 mg/L of dissolved salts.

Only moderate agricultural activity has developed in the area and the catchment area is sparsely inhabited, with almost no industry. Therefore, the water quality of the wadi is still good.

A dam was constructed in the eastern, upstream part of the wadi to collect good quality base and flood flows from the area covered by calcareous and basaltic rocks with low salinity of about 500 µS/cm. Water is released from the dam to flow along Wadi Mujib where it is joined by spring water of a lower quality with an average salinity of 1200 µS/cm. At the wadi's entrance into the Dead Sea area its water is captured and pumped for municipal use.

Wadi El-Karak

The catchment area of Wadi El-Karak measures 190 km² and lies at elevations ranging from 1000 masl along the highlands to 400 mbsl at the shores of the Dead Sea. The average precipitation falling over the catchment area ranges between 350 mm/year in the mountains to 100 mm/year along the shores of the Dead Sea. The

Fig. 2.13 Catchment area of Wadi Mujib

potential evaporation rates range from 3100 mm/year along the highlands to 2600 mm/year at the shores of the Dead Sea.

This catchment area is moderately inhabited, agrarian and includes the city of Karak and numerous towns and villages. Karak City and surrounding villages are sewered, the waste water is treated and the effluents are discharged into Wadi Karak.

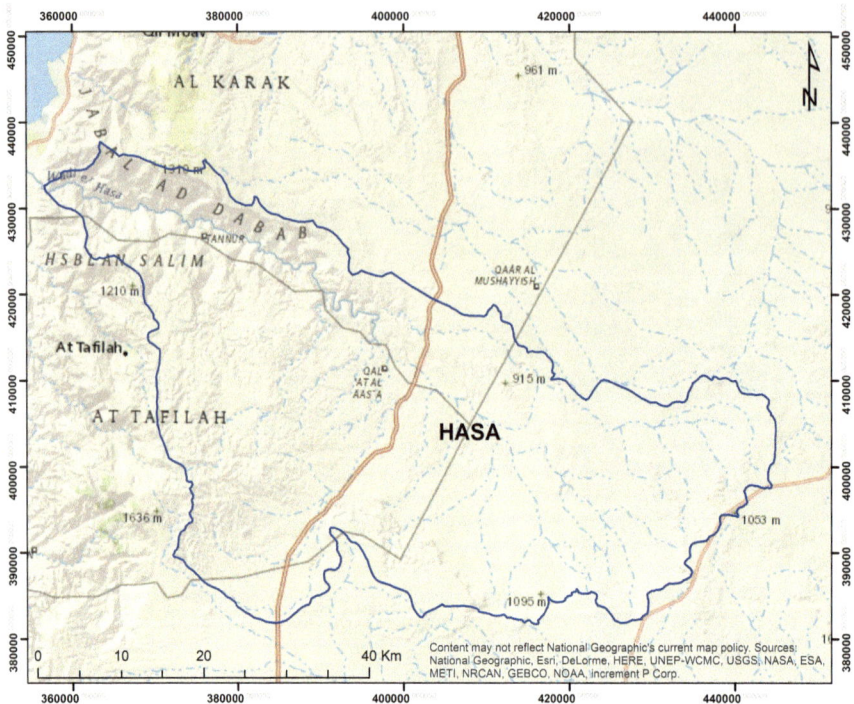

Fig. 2.14   Catchment area of Wadi Hasa

All wadis east of the Dead Sea including Wadi El-Karak are rich in springs and water seepages issuing from the sandstone aquifers.

The average base flow discharge of Wadi El-Karak is around 15 MCM/year and the flood flow is around 3 MCM/year. The base flow is used for irrigation along the course of the wadi. The salinity of the base flow in the downstream reaches of the wadi measures around 1000 μS/cm and it increases towards the Dead Sea to reach many thousand μS/cm (Fig. 2.14).

Wadis Between the Major Dead Sea Catchments
Different small areas (inter-catchments) between the major Dead Sea catchments discharge directly into the Dead Sea such as the catchment areas between Wadi Hisban and Zarqa Ma'in, Zarqa Ma'in and Mujib, Mujib and Karak, and Karak and Hasa. Their total catchment areas measure 972 km^2, with a total discharge of around 30 MCM/year. This discharge mostly originates from groundwater issuing along the lower reaches of the wadis as thermal mineralized water with salinities ranging from 500 μS/cm in the area between Wadis Hasa and Karak to several thousand μS/cm between Wadis Hisban and Zarqa Ma'in, also increasing very strongly from south to north.

2.1.3   Wadi Araba Catchments

Wadi Araba itself does not form a base level for surface or groundwater. The surface and groundwater of its northern part discharge into the Dead Sea and that of its southern part discharges into the Red Sea via the Gulf of Aqaba.

Northern Wadi Araba Catchment
Wadi Araba catchment has a length of 200 km, a width of 25–30 km and a total area of 2938 km^2. Its northern part extends 100 km from the Dead Sea shore southward. Precipitation over the highlands averages 300 mm/year falling partly as snow. The average long-term precipitation in Wadi Araba area itself is around 100 mm/year. The potential evaporation rates range from 2800 mm/year at the southern shores of the Dead Sea to 3500 mm/year in the semi-arid south-eastern parts of the catchment.

Different wadis drain the catchment area into Wadi Araba. Major among these are Wadi Khuneizir, Wadi Fidan and Wadi Buweirida, with average discharges of about 11.4, 5.5 and 3 MCM/year, respectively. The major part of the discharge consists of the base flow of wadis.

In addition to the major wadi catchments, numerous small inter-catchments drain the area. The overall total discharge of all the northern wadis into Wadi Araba is 26 MCM/year.

Course-grained alluvium deposits build up on the bottom of Wadi Araba; therefore, flood flows that reach the wadi infiltrate rapidly to recharge the groundwater. They seldom reach the Dead Sea directly, but the infiltrated water flows as groundwater in a northerly direction towards the Dead Sea and discharges as seepages or submarine springs into the Dead Sea.

The catchment area is sparsely populated and the main centers of Tafilah and Shoubak are devoid of major industry. Agricultural activity takes place in the highlands, where rain-fed crops are produced, and also along the side wadi courses and in Wadi Araba. Here, base flows of wadis and groundwater are used for irrigation.

Tafilah Town is sewered, the waste water is treated in a waste water treatment plant and the effluents are discharged and used in irrigation along Wadi Fifa. The amount of effluent is very small (a few hundred cubic meters per day); part of it infiltrates along Wadi Fifa and discharges with the groundwater issuing along the lower reaches of that wadi, or it joins the groundwater in Wadi Araba.

Abu Zirr (1989) calculated the natural groundwater throughput of northern Wadi Araba into the Dead Sea to average 22 MCM/year. Well drilling and agricultural development increased in the Wadi Araba area after the peace treaty with Israel in 1994, and hence the groundwater amounts reaching the Dead Sea have since declined. At this time the decline cannot be estimated because the groundwater flow regime is in a transition phase.

Southern Wadi Araba Catchment
The catchment area measures 1278 km^2 and extends from Aqaba northward to around 100 km with an E-W width of 30 km. The average precipitation is 150 mm/

year in the north-eastern parts decreasing to less than 50 mm/year in the southern parts and Aqaba area. The potential evaporation rates range from 3300 mm/year in the northern parts to 4100 mm/year in the southern parts.

The area is barren, with a very low population density of less than 1 person/km². The aridity of the area does not support life and does not allow easy urbanization.

The very low water potentialities of the area are indicated by the estimated total water discharge from the eastern mountain wadis of 1 MCM/year.

The groundwater coming from the eastern highland's aquifers into Wadi Araba alluvial aquifer flows as aquifer to aquifer lateral discharge and flows through the alluvial deposits to the Gulf of Aqaba. The output of groundwater of southern Wadi Araba into the Red Sea is around 10 MCM/year.

2.1.4   Wadi Yutum Catchment

Wadi Yutum catchment of 4.400 km² drains an extensive area in south-west Jordan, east of Aqaba into the Red Sea. Precipitation over the area ranges from 150 mm/year in the highlands to less than 50 mm/year in the central and eastern parts of the catchment area. The potential evaporation rates range from 3400 mm/year in the western highlands up to 3800 mm/year in the eastern and southern flat areas.

Since most of the area is flat and comprised of friable sediments possessing high porosity and permeability, precipitation water infiltrates rapidly into these rocks and recharges the groundwater. When the infiltration capacity of the soil cover is exceeded as a result of intense precipitation, powerful floods occur and cause damage to infrastructures along the wadi courses, although the amounts of water involved are relatively small of about 1.5 MCM/year as related to the extent of the catchment area.

2.1.5   Jafr Basin Catchment

Jafr basin is an exitless depression in southern Jordan with a catchment area of 12.200 km² (Fig. 2.15). The larger part of the area is flat and only a small western part of it is comprised of highlands. The average precipitation rates range from 200 mm/year in the highlands to 30 mm/year in the middle and eastern parts of the catchment and the potential evaporation ranges from 3300 mm/year in the western highlands to 4000 mm/year in the central and eastern depression.

The flood flow of the catchment is around 10 MCM/year and the base flow is 5 MCM/year. Both flows collect in the Jafr depression, where they either evaporate or infiltrate into the groundwater there. At present, the base flow of spring discharge is entirely used in irrigation.

The catchment area is very sparsely populated, with Ma'an and Shoubak as the major urban centers. Agriculture entirely along the foothills of the mountains in the

Fig. 2.15 Catchment area of Jafr basin

west by extracting groundwater. The main industry in the area is a cement factory located at the north-western edge of the catchment in addition to entirely extraction of stone for building and sand for glass.

2.1.6 *Azraq Basin Catchment*

The center of Azraq basin lies at an altitude of 500 masl creating an exitlesst opographic depression in the eastern plateau of Jordan in which an oasis has formed (Fig. 2.16). The drainage basin measures 11.600 km² and extends in the north beyond the borders of Jordan into Syria.

Precipitation over the area ranges from 300 mm/year over the southern slopes of Jabel Druz to less than 50 mm/year in the Azraq depression itself with an average precipitation over the catchment area of 90 mm/year, some of which is received as snowfall in the north-western parts of the catchment.

The potential evaporation ranges from 3300 mm/year in the northern parts of the catchment to 4000 mm/year in its central and eastern parts.

The total discharge of the catchment area is around 27 MCM/year, of which 15 MCM/year issues as groundwater from different springs in the Azraq Oasis

Fig. 2.16 Catchment area of Azraq basin

itself. The rest consists of flood water originating from precipitation events over the catchment and pouring along wadis into the depression.

Only a few dispersed urban centers and small types of industries are found in the catchment area. For the last three decades or so the spring water feeding Azraq Oasis has been pumped to the urban centers of Amman and Irbid, causing the drying of the springs at the end of the last century.

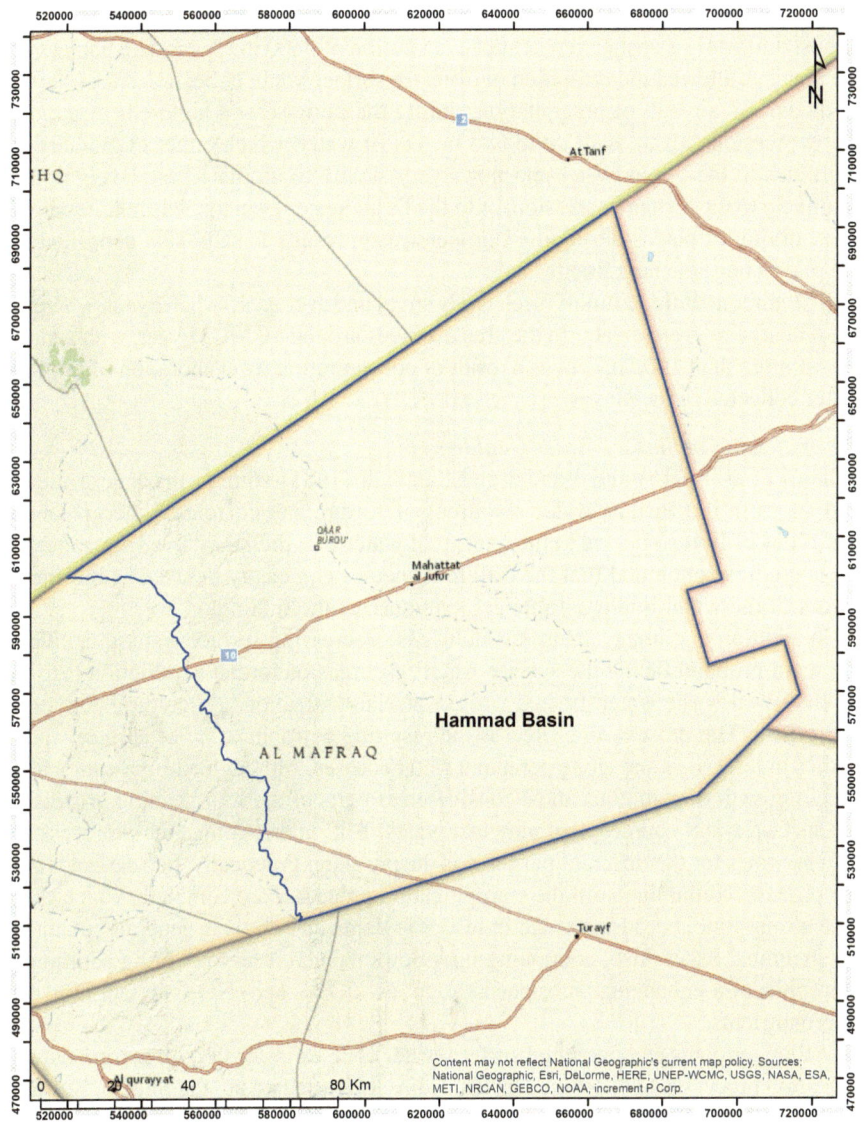

Fig. 2.17 Catchment area of Hammad basin

2.1.7   *Hammad Basin Catchment*

Hammad basin is a very large, flat plateau extending into four countries: Jordan, Syria, Iraq and Saudi Arabia, with an area in Jordan measuring 19.270 km² (Fig. 2.17).

Precipitation rates over the area range from 150 mm/year to 50 mm/year with almost uniformly distributed potential evaporation of 3800 mm/year. The flatness of the area has allowed the formation of different surface water collection sites (playas or qa'as). As a result of precipitation, during the rainy season hundreds of almost flat depressions fill up with up to two meters of water, which either evaporates or infiltrates to the groundwater and flows very slowly to ultimate base levels in the south; Go to the Sirhan depression, or to the Dead Sea or even to the Euphrates area. They drying of playas during the summer season results in salty silts deposited in the playas until the next flood.

The amount of flood runoff is relatively small and averages 5 MCM/year, whereas the recharge to groundwater in the area averages around 10 MCM/year, with salinities of more than 1000 µS/cm as a result of concentration by evaporation of rainfall water collected in the playas.

Bio-Indicators of Surface Water Qualities
Alhejoj et al. (2015) and Bandel and Salameh (1981) studied the macro-fauna and -flora in the surface water resources of Jordan and correlated the presence and types of living species to the quality of water in which they lived. In general, these studies concluded that the different species can easily be used as bio-indicators of the water quality. Table 2.1 summarizes the findings.

In addition to climate changes which cause decreasing surface water quantities, the main problem facing the surface water resources of Jordan is pollution caused by industrial waste water, treated waste water, urbanization, agricultural activities and others. The most severe effect is the resulting eutrophication of surface water bodies in reservoirs and along perennial water courses. Although eutrophication and access growth of plankton can be considered as parts of the self-cleaning processes for surface water sources, this may take years. But, interrupting such a process to use the water for the different purposes is the problem. Practically, one cannot leave the process to continue until the eutrophication processes are complete, which may take a long time, because storage in reservoir dams and pools is generally planned on an annual basis. Also, continuing inlet flows in such reservoirs bring additional eutrophication-enhancing substances such as PO_4^{-3} and NO_3^- in addition to micronutrients.

All major dams and perennial surface water courses in Jordan suffer from eutrophication processes to different degrees. Dams thus affected are the Shueib, Kafrain and King Talal dams, and the wadis are Zarqa, Shueib, Kafrain Abu Nuseir whose waters are hyper-eutrophic; whereas the Wahda (Unity), Wadi Arab, Ziglab and Kufranja dams, and the rivers and wadis: Yarmouk, Ziglab, Arab, Yabis are eutrophic; and dams of Wala, Mujib and Tannur are mesotrophic. This is due to the natural presence of NO_3^- in rain and flood water of about 12 mg/L and the leaching of phosphate rocks covering large areas in the country, augmented by human activities that produce substances leading to eutrophication.

Table 2.1 Jordanian Bio-monitor System for Watercourses (JBSW) which classifies water qualities in 12 categories (Alhejoj et al. 2014)

	Water quality	Aquatic organisms bio-indicators	Tolerance degree	Environment description including; substrates, water body type, and color.	Chemical indicators	Locations, examples
1	High water quality (very clean/ non-polluted)	Theodoxus, Turbellarian (flat worm), and Mayfly larvae (*Caenis antoninae*)	Organisms highly sensitive to pollution	Rocks and gravels, clear shallow water from still to slow running water	Low COD, BOD values and main ions;Ca^{2+} and Mg^{2+}	Hisban and Shita Springs
2	Very good water quality by self- purification	No Turbellarian (flatworm). Theodoxus, Melanopsis, Mayfly larvae, (*Baetis monnerati*), and sponges	Organisms highly sensitive to pollution and not sensitive to strong growth of phytoplankton (no nitrate)	*Exposed* riverbed *rocks* but with darker mud deposits than normal due to additional nitrate. Shallow water from moderate to swift current	COD value is around 15 mg/L and BOD 0 mg/L. High values of SO_4 (>7 meq/L), NO_3 about 0.2 meq/L, and Na 5 meq/L	Mujib River a few km below dam site and KAC during spring 2013
3	Good water quality	Gastropods (Melanopsis, Galba and Bulinus). Crustacea (Amphipoda) and Mayfly larvae such as *Baetis monnerati*	Organisms sensitive to pollution, clean water to slightly polluted water	Gravel bottom stream with moderate running water and springs issuing from carbonate rocks	Normal freshwater chemistry but small increase in NO_3 and PO_4 values	W. Hisban below spring area, Wadi Sir, small spring creek in upper Wadi Zarqa
4	Fairly good water quality	Melanopsis coming from continuous sources, Physa, Mayfly larvae (Caenis sp.), and leeches	Semi-tolerant organisms that tolerate some degree of house sewerage added to streams	Rock and gravel, less sand occurs in the relatively turbid running water from moderate to swift current with high occurrences of aquatic vegetation	NO_3 concentration up to 1.0 meq/L and low concentration of Na dissolved ions	W. Shueib and Wadi Sir upstream of the treatment plant

(continued)

Table 2.1 (continued)

	Water quality	Aquatic organisms bio-indicators	Tolerance degree	Environment description including; substrates, water body type, and color.	Chemical indicators	Locations, examples
5	Moderate	Chironomidae (blood worms), Physa, Mayfly larvae, and Caddisflies larvae	Tolerant organisms which survive with moderately polluted water	Bottom surface between stones with organic mud and sand below surface black	COD value reach to 50 mg/L and BOD 22 mg/L. SO_4^{2-} around 6 meq/L, NO_3^- 1.5 meq/L, and PO_4 0.3 mg/L	Periodically in lower Zarqa River
6	Slightly bad	Low densities of Physa and leeches or disappearance and well present Simulidae (Black fly larvae) and tube-building animals such as Chironomid (blood worms), and Tubifex (sludge worms)	Pollution tolerant organisms but still with well aerated water	Rocks with muddy organic rich deposits in interspaces. Brown, turbid water with moderate to swift current	Relatively high values of COD with 60 mg/L and BOD reach 35 mg/L. SO_4^{2-} about 1,5 meq/L, NO_3^- 1.0 meq/L, and PO_4 1.0 mg/L	Wadi Shueib downstream of the treatment plant and periodically parts of Zarqa River
7	Bad (Eutrophic)	High occurrence of Chironomidae (blood worms), Tubifex and Physa with their eggs	Tolerant to eutrophication organisms	Fine sediment below surface and lower side of rocks black with still greenish turbid water and high growth of algae and Cyanobacteria	COD is 35.0 mg/L and BOD 5.0 mg/L.SO_4^{2-} (4.5 meq/L),NO_3^-(0.20 meq/L), and PO_4 (0.30 mg/L)	Wadi Mujib Lake directly below dam
8	Very bad water quality	Red Chironomid worms, frogs and fish	Organisms not sensitive to pollution	Muddy sediment between rocks, black below surface, rocks covered with bacterial slime	COD >70 mg/L and BOD >10 mg/L. NO_3^- (1.50 meq/L), PO_4 (0.30 mg/L), HCO_3^- (8.0 meq/L), and CL (11.98 meq/L)	Periodically Zarqa River especially upstream of king Talal reservoir.

9	Poor water quality (Heavily polluted water) (smelly and colored water)	No insect larvae only bacterial slimes. In Kafrain lake fish periodically killed	Organisms highly tolerant to pollution	Muddy with high sludge content. Turbid – gray color in creeks brown in lake	COD up to 83.0 mg/L and BOD 40.0 mg/L. high concentration of PO_4 and NO_3	Wadi Sir after and below treatment plant and Kafrain dam
10	Mineral water	Large numbers of Pseudamicola (snails), Simulidae (Black fly larvae) and Chironomid worms. Larvae of mayflies such as *Nigrobaetis vuatazi* and caddisflies. Springs with mosses and ferns	Organisms tolerant to mineral water and slightly salty water	Springs with gravel and silty ground in creeks	Relatively high concentrations of Sr (reaching 37.0 mg/L) Br 10.0 mg/L, and Zn 0.02 mg/L	Wadi Atun stream
11	Warm water	No macro-aquatic animals only dense crust of Cyanobacteria and springs with reeds (Phragmites)	Organisms are able to survive high temperatures	Warm springs issuing from basalt and sandstone as warm as 60 °C	Relatively high concentration of trace elements such as Mn 5.0 mg/L, Cu 0.2 mg/L, Ba 700 mg/L	Zara and Zarqa-Ma'in springs
12	Brackish water	Class 1	Pseudamnicola (snails), Ostracods (e.g., *Heterocypris salina*, *Cyprideis torosa*, and *Heterocypris reptans*) and insect larvae such as Ephydridae (shore fly), Chironomidae, Mayfly (*Cloeon* sp.).	Salt tolerant organisms which survive with salinities of 16,000–21,000 µS/cm	Creek on gravel with algae and aquatic plants. Pools surrounded by salt tolerant plants with ostracodes	Relatively high concentrations of dissolved salts. Br and Sr	Karama area
		Class 2	Periodically with ostracods (e.g., *Heterocypris reptans* and *Cyprideis torosa*) up to 2011 with Melanoides but from 2012 on without Melanoides- surrounded by salt tolerant plants	Up to 2011 with salinity about 23,600 µS/cm. In 2012 and 2013 the salinity level reached 26,100 µS/cm	Standing water with muddy ground	Higher salt concentrations and Br 16 mg/L and Sr 50 mg/L	Karama reservoir lake.

2.2   Groundwater

Groundwater Aquifers

The groundwater in Jordan is found in three main aquifer complexes (Figs. 2.18 and 2.19).

– The Deep Sandstone Aquifer
– The Upper Cretaceous Aquifer and
– The Shallow Aquifer.

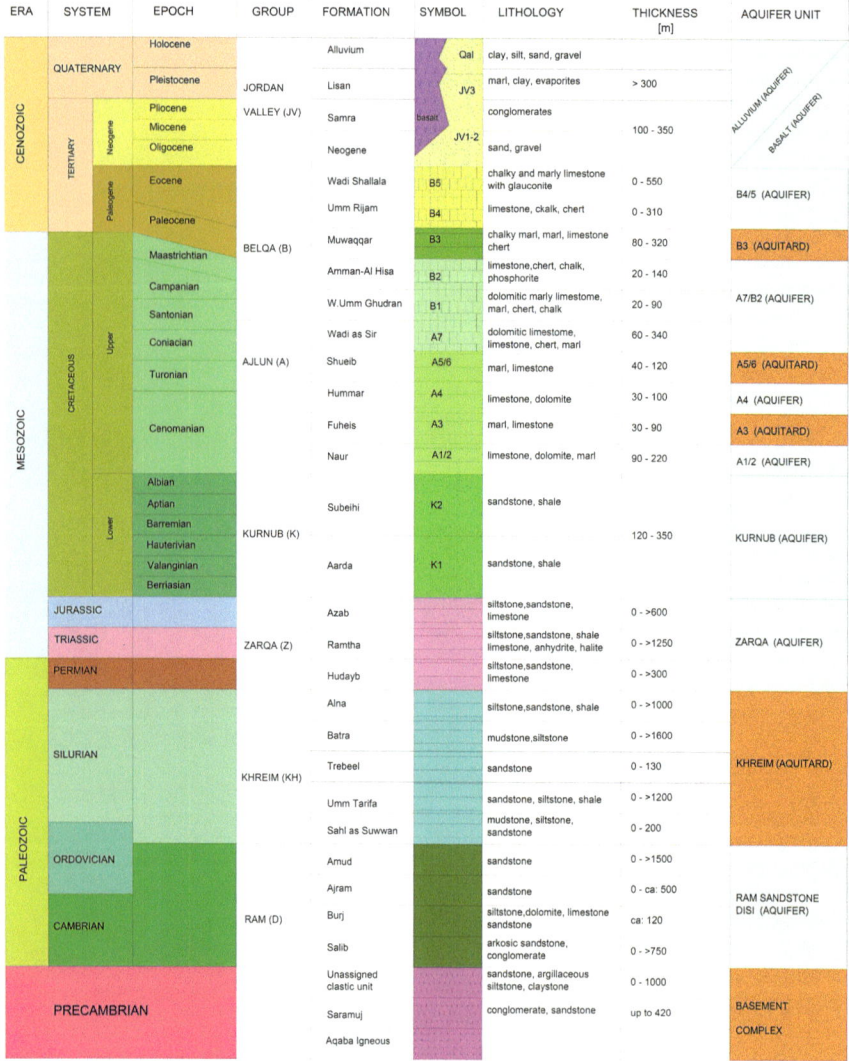

Fig. 2.18 Geologic column of Jordan. Lithologies, ages, thicknesses and hydraulic properties (NRA open files)

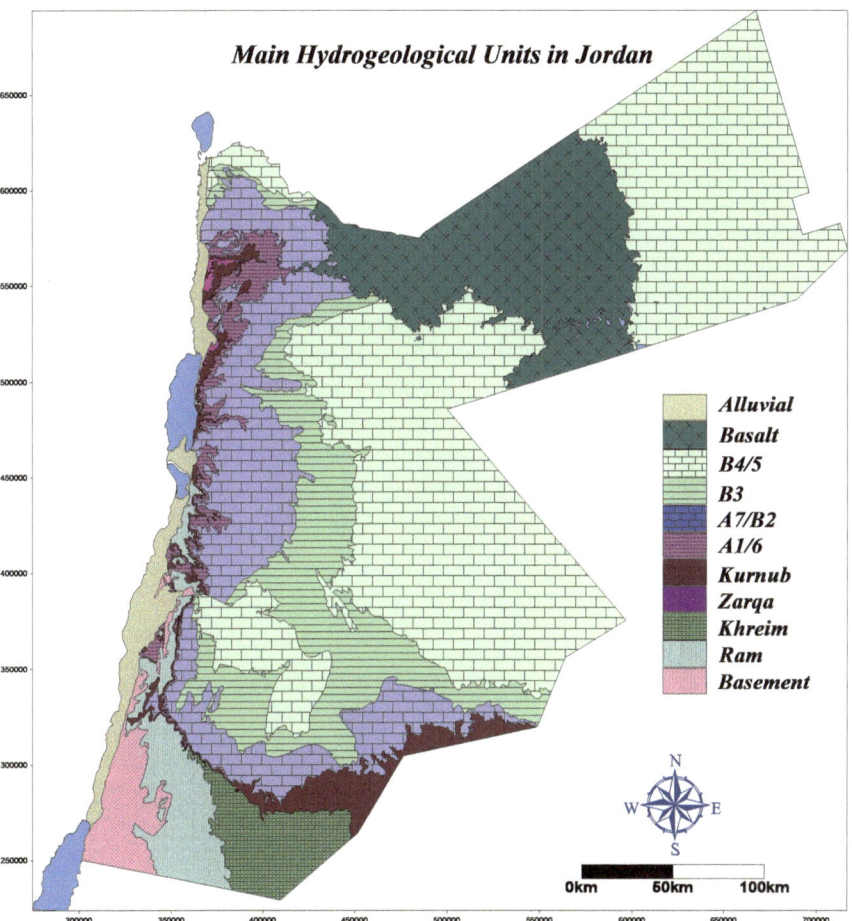

Fig. 2.19 Surface distribution of the main hydrogeological units in Jordan (MWI open files)

2.2.1 Deep Sandstone Aquifer Complex

This complex underlies the whole territory of Jordan as one geologic unit. In north Jordan it gradually becomes separated into two aquifer parts by the intercalation of thick limestone, silt and marl formations of Triassic and Jurassic rocks forming the pre-Triassic and post-Jurassic deep sandstone aquifer complex parts but which, nonetheless, remain hydraulically interconnected.

Disi Group Aquifer (Paleozoic)
This is the oldest, and in the north, the deepest water bearing sediment sequence in Jordan, consisting of sandstones and quartzites overlying the granitic basement complex of the Arabian Shield. It crops out only in the southern part of Jordan and partly along the Rift Valley. The southern part of the complex forms the fresh water aquifer of the southern desert area of Jordan.

The groundwater flow in this aquifer complex is directed towards central Jordan lying further north-east and from where it gradually changes its flow direction to north-west and west where it then discharges along the Dead Sea escarpment (Fig. 2.20). Only in the southern parts of the Disi Rum area a groundwater divide separates a small southern region where the groundwater moves directly towards the west and south towards Wadi Araba-Aqaba areas.

Kurnub and Zarqa Group of Triassic-Lower Cretaceous Age
This is mainly a sandstone aquifer underlying the area of Jordan and overlying the Disi group aquifer in southern Jordan. In north Jordan it becomes separated from the underlying Disi aquifer complex by Triassic and Jurassic formations. It crops out along the lower reaches of the Zarqa River basin, in Baqa'a area and along the escarpments of Ras en Naqab and the southern Jordan Rift Valley area.

Wells drilled in this aquifer have good productivities. Direct recharge takes place in the small outcrop areas where the groundwater has low salinities; otherwise the groundwater in this aquifer is mineralized.

The Kurnub-Zarqa aquifer system is being exploited mainly in the lower Zarqa River and in and Baqa'a areas.

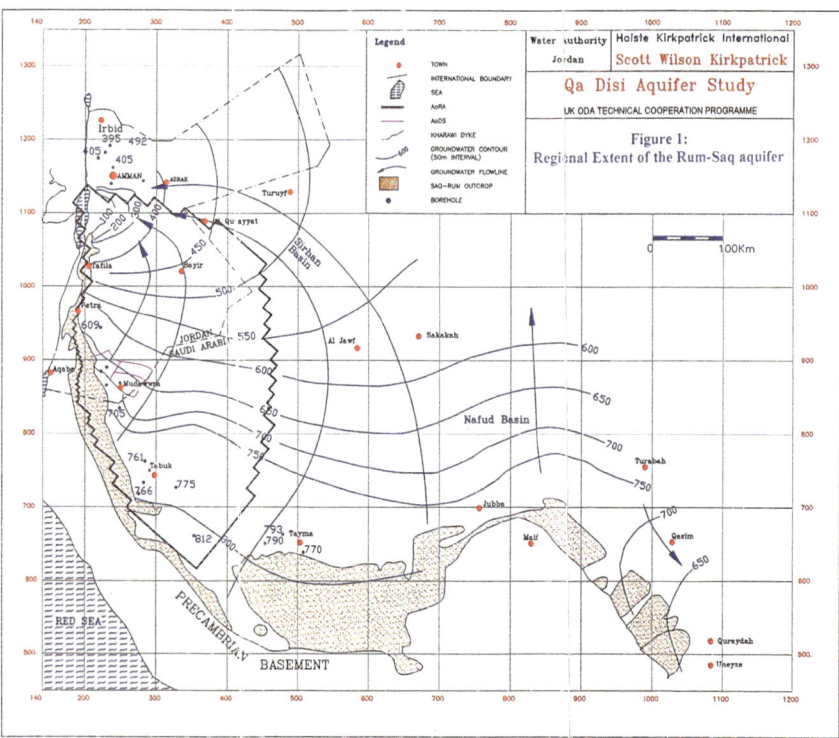

Fig. 2.20 Groundwater flow pattern in Disi (Ram-Saq) aquifer complex

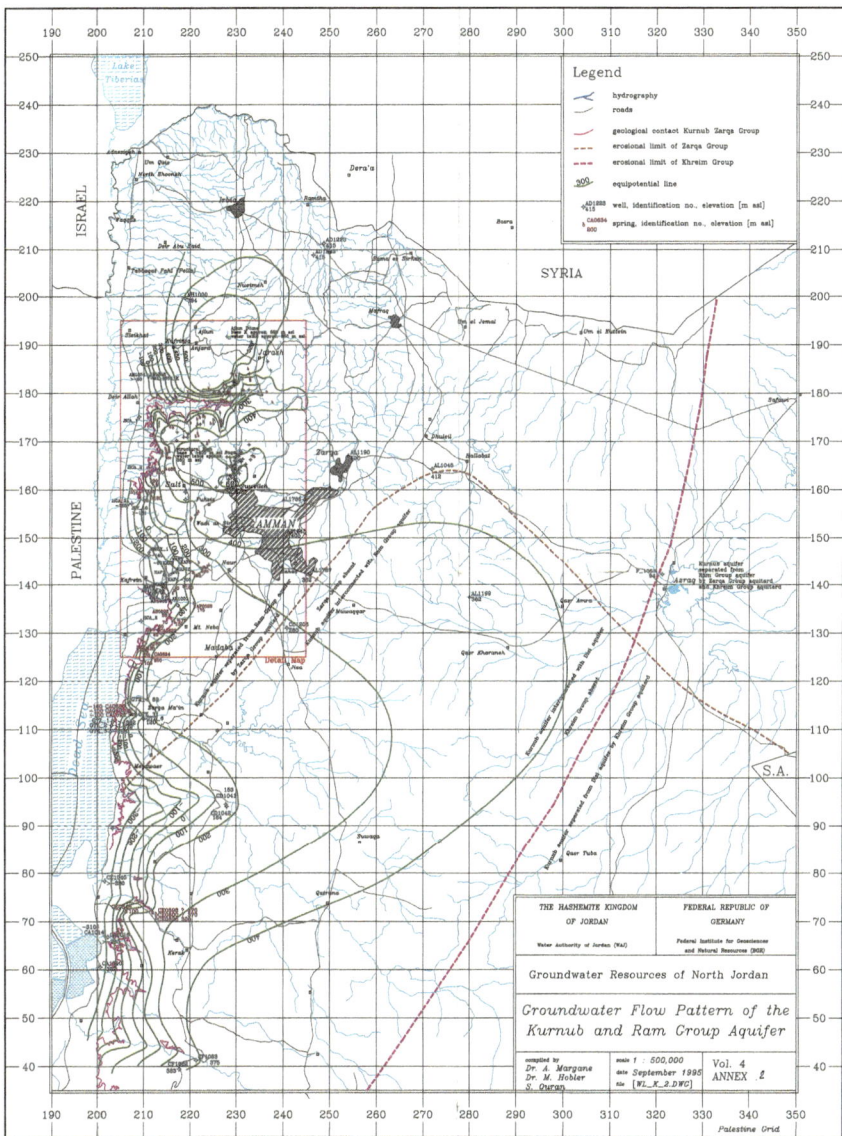

Fig. 2.21 Groundwater flow pattern of the lower cretaceous aquifer (Kurnub). To the south of latitude 110 the Ram aquifer directly underlies the Kurnub aquifer to form one aquifer complex designated as the deep aquifer complex (BGR 1995)

The direction of groundwater flow in this aquifer system is generally directed to the ultimate base level; the Dead Sea: towards north-east in the southern part of Jordan, towards west in central Jordan and towards south-west in northern Jordan (BGR 1995; Salameh and Udluft 1985), (Fig. 2.21).

2.2.2 Upper Cretaceous Hydraulic Complex

Alternating sequences of limestone, dolomite, marlstone and chert beds build this aquifer complex with a total thickness in central Jordan of about 700 m. Its limestone, dolomite and chert units form excellent aquifers.

The Na'ur Formation of this complex (A1/2) consists of about 200 m of marls and limestone and possesses in some areas relatively high permeability forming a potential aquifer. The overlying aquitards; Fuheis Formation (A3) consists of about 80 m of marl and shale and overlies the A1/2, separating it from the A3-overlying Hummar Formation aquifer (A4). The A4 consists of massive, pure, semi-crystalline and karstic limestone entailing its very high permeability and porosity. The outcrops of the A4 aquifer along the highlands allow its recharge by precipitation water. To the east the aquifer becomes confined by the overlying Shueib Formation composed of impermeable alternations of marl and limestone beds (A5/6).

The A5/6 aquitard is overlain by the Massive Silicified Limestone Formation (A7/B2), which consists of limestone, chert-limestone, sandy limestone, marly limestone and phosphatic beds building the most important aquifer of the complex, recharged along the highland where it crops out and receives direct recharge.

In the eastern parts of the country this aquifer complex is overlain by the thick marl and shale beds of the Bituminous Muwaqqar Formation (B3), forming a competent confining formation. Therefore, in some locations, flowing artesian wells are drilled into the A7/B2 aquifer (Fig. 2.22).

The groundwater flow in this complex is directed from the highlands, where recharge takes place, partly to the escarpment of the Jordan Rift Valley within the faulted blocks or to the eastern desert, where it discharges along deeply incised wadis or it flows further eastwards. Along its flow paths to the east, part of the groundwater seeps down through the fractured and jointed series of aquifers and aquicludes (A5/6 to A1) into the underlying sandstone aquifer complex, and the other part discharges from springs and seepages in depressions like Azraq and Sirhan (Fig. 2.23).

Water over the highlands infiltrates into the outcropping Upper Cretaceous aquifers and moves either westwards towards the Jordan Rift Valley or eastward towards the desert areas where it infiltrates deeper into the Lower Cretaceous and Palaeozoic aquifers. Here it then flows westwards towards the Dead Sea, which is the ultimate base level of all surface and groundwater bodies of Jordan (Salameh and Udluft 1985).

2.2.3 Shallow Aquifers Hydraulic Complex

The shallow aquifers hydraulic complex consists of three main systems:

Rijam and Shallala Formations
Rijam and Shallala Formations consist of chalk marls with a thickness up to 800 m deposited during the Eocene. They form the shallow aquifer in the eastern parts of

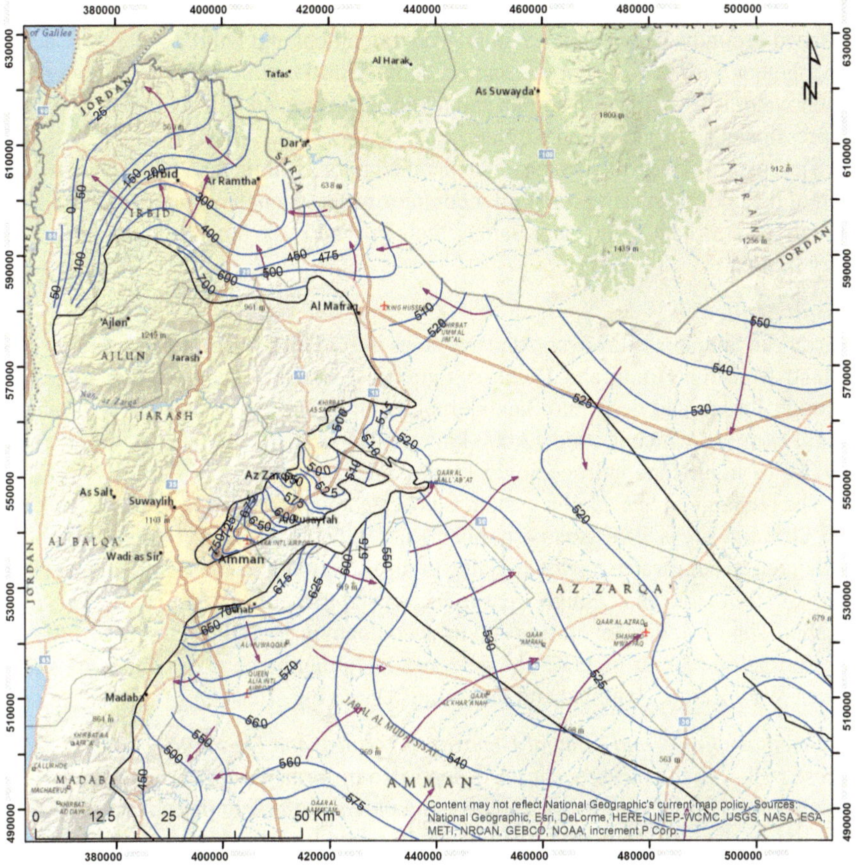

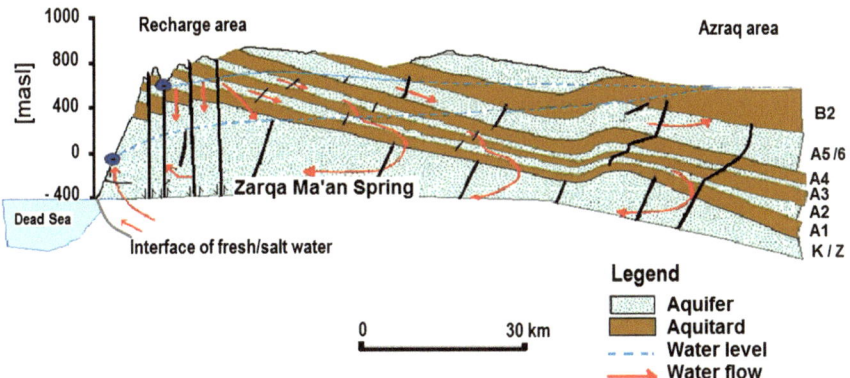

Fig. 2.22 Groundwater flow B2A7

Fig. 2.23 Groundwater flow pattern on the eastern side of the Dead Sea recharge. (Modified from Salameh and Udluft 1985)

the country especially in the Jafr and Sirhan-Azraq depressions. They contain perched groundwater in the Yarmouk River catchment and build a composite aquifer together with the overlying basalts of Jabal Arab-Druz in Harrat ash Sham area. Their water is fresh in north Jordan and in the basalt area and fresh to brackish in the depressions of Jafr and Sirhan-Azraq.

The quantity of groundwater in the perched parts is limited and is considered as part of the underlying shallow aquifer complex water when they underlie the basalts. In Jafr and Sirhan-Azraq depressions their groundwater quantities are generally very limited.

Basalt Aquifer
Harrat esh Sham basalts cover extensive areas in Jordan, Syria and Saudi Arabia and extend from the Huran and Damascus area in Syria in a south-eastern direction through Jordan and deep into Saudi Arabia with a total outcropping area of about 45,000 km^2, of which around 12,000 km^2 are in Jordan. These basalts form a good aquifer of significant hydrogeological importance.

The recharge to this aquifer system is provided by precipitation, especially along the elevated areas of Harrat esh Sham and Jabal Arab-Druz, from where the recharged groundwater moves radially in all directions; to the Damascus area in Syria, to the Azraq and Dhuleil areas in Jordan and to the Sirhan depression in Saudi Arabia. The geological evolution and its accompanying geologic structures favored the formation of four main discharge zones in Jordan namely, the upper Yarmouk River, Wadi Zarqa, Hammad and the Azraq basins (Fig. 2.24).

Sedimentary Rocks and Alluvial Deposits of Tertiary and Quaternary Ages
These rocks are the most recent rocks deposited in Jordan. They form local aquifers in most parts of the country partly overlying older aquifers. In places they are separated from them by aquitards. Tertiary aquifers are found in depressions along the highlands such as those in the surroundings of Jafr and Azraq and in the Jordan Rift Valley area. Quaternary aquifers are mainly concentrated in the eastern desert, Jordan Rift Valley area, Jafr and Azraq basins and along river and wadi courses (Fig. 2.25).

Recharge into these aquifers takes place directly from precipitation and flood water or through the overlying basalt aquifer, as in the case of Azraq basin, or from the surrounding aquifers, as in the cases of the Jordan Rift Valley and Aqaba areas.

The groundwater flow in these aquifers in the highlands is directed radially towards the Azraq and El-Jafr depressions.

The groundwater flow in these sediment fills is highly dependent on the underground conditions, but it is mainly oriented towards wadi courses.

2.2.4 Groundwater Basins in Jordan

The groundwater basins in Jordan (Fig. 2.26) were separated from each other according to the principle that groundwater basins or groundwater balance areas are those areas which could be separated and defined to include appropriate and

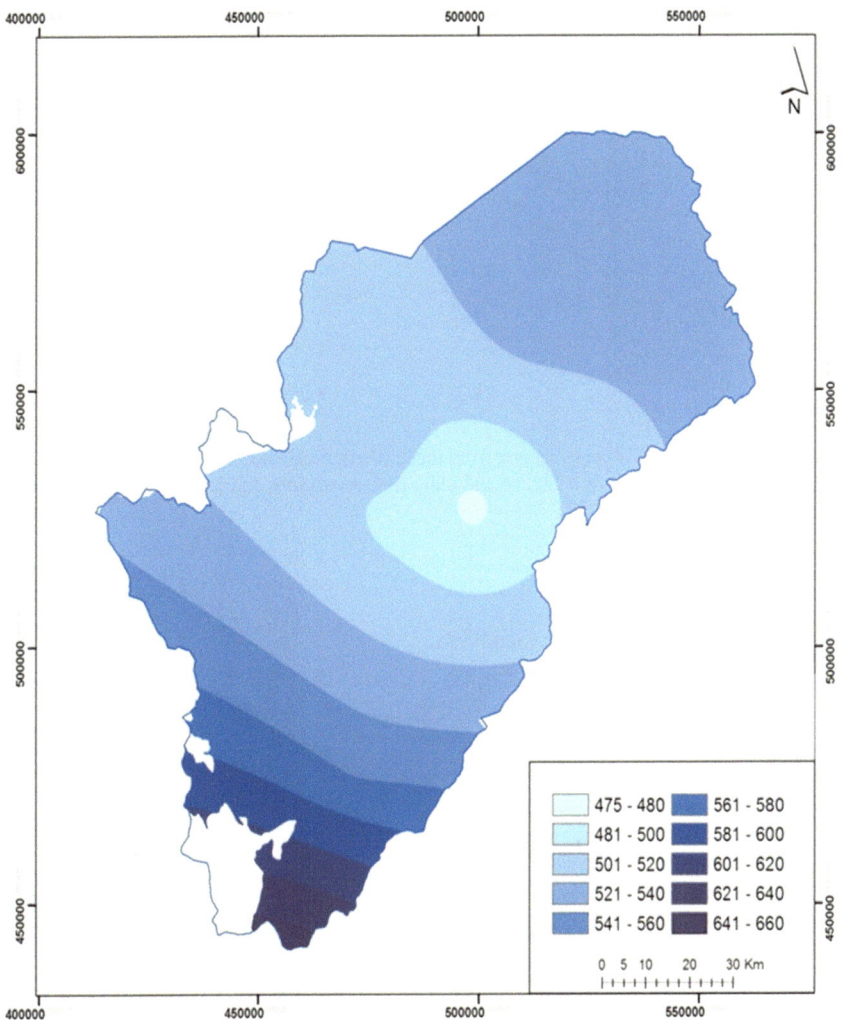

Fig. 2.24 Groundwater flow pattern in the Alluvium, Basalt and B4/B5 aquifers

regionally important aquifer systems. Some of the recharge and discharge takes place within the same basin, but generally groundwater movement between basins is the norm. A groundwater basin can contain more than one aquifer or aquifer complex or system; therefore, a groundwater basin should be related to a certain aquifer system and not to all aquifer systems underlying the basin.

The National Water Master Plan of Jordan (NWMP 1977) defined the groundwater basins in Jordan as listed below, but, in this work Jafr is subdivided into Jafr and Disi-Mudawwara basins:

1. Yarmouk
2. Northern escarpment to the Jordan Valley

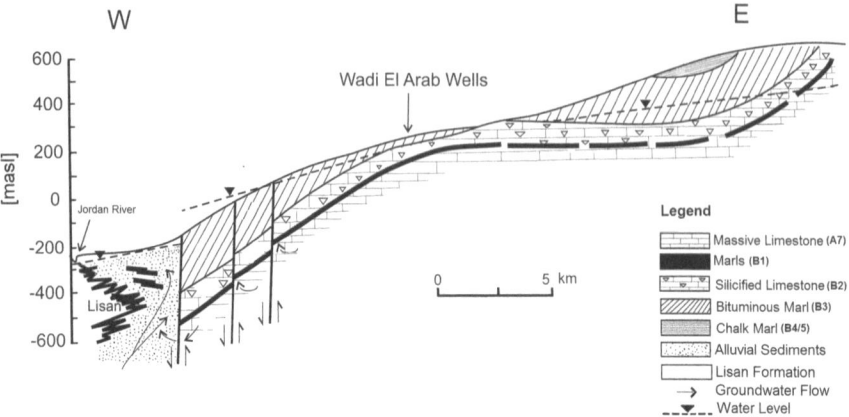

Fig. 2.25 Geological E-W cross section from the northern highlands to the Jordan River showing the geological constellation of Neogene and older rock formations and recent alluvial deposits of the Jordan Valley

3. Jordan Valley floor
4. Zarqa River
5. Central escarpment to the Dead Sea
6. West Bank
7. Escarpment to Wadi Araba
8. Red Sea
9. Jafr
10. Azraq
11. Sirhan
12. Wadi Hammad
13. Disi-Mudawwara.

2.2.4.1 Groundwater in the Yarmouk Basin and the Northern Part of the Jordan Valley Escarpment

The main groundwater body in the Yarmouk River basin is found in the B2/A7 aquifer at depths of less than 200 m in the highlands. The groundwater in this aquifer flows towards the Jordan Valley and Yarmouk River areas. The marly Muwaqqar Formation B3 forms an aquiclude overlying the Amman Wadi Sir aquifer system B2/A7 all dipping towards the Yarmouk and Jordan Rivers. The B2/A7 aquifer becomes gradually confined in a westerly direction towards the areas of the Jordan and Yarmouk Rivers. Wells drilled in this aquifer along the mountain slopes to the Jordan Valley and the Yarmouk River encountered confined or even artesian water with piezometric heads measuring tens of meters above the ground surface.

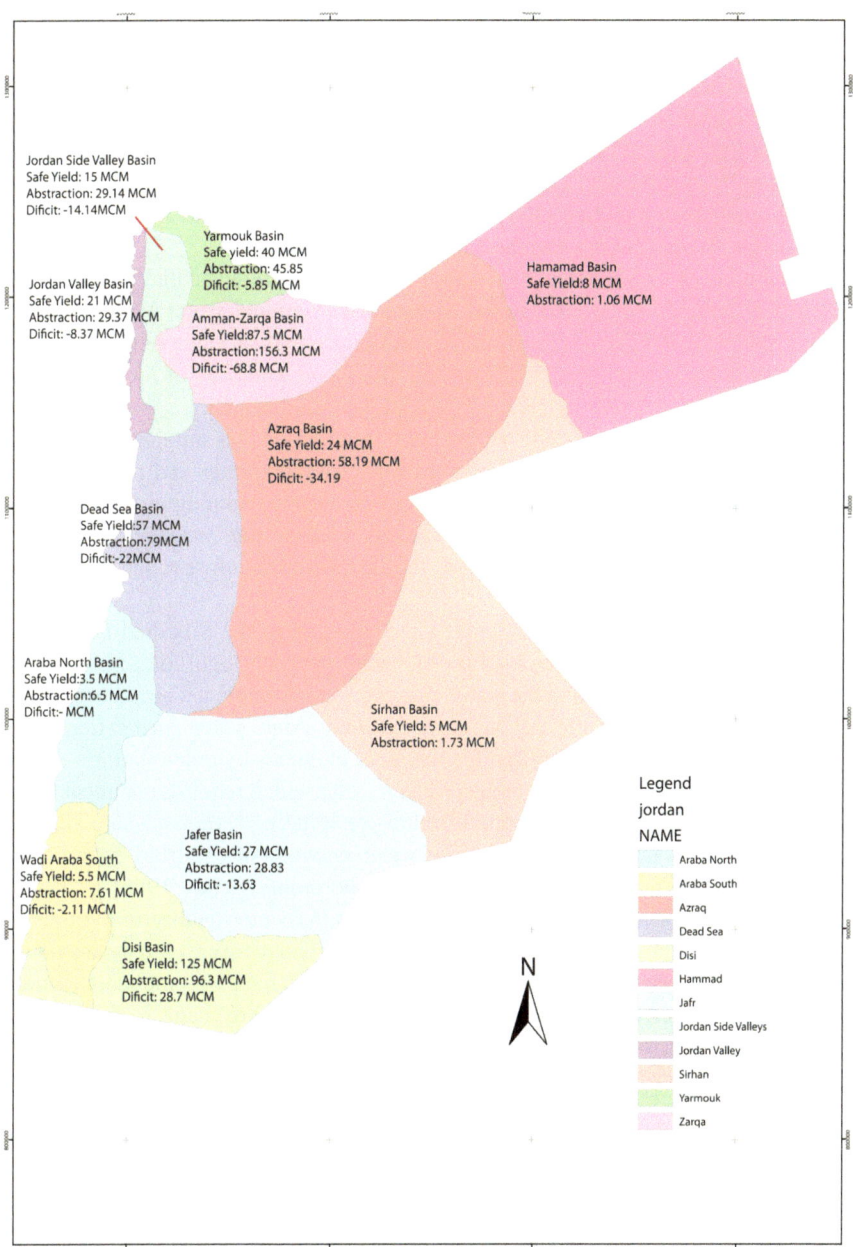

Fig. 2.26 Groundwater basins in Jordan (MWI open files)

Mukheiba wells at the slopes to the Yarmouk River and Wadi El-Arab wells in Wadi El-Arab, upstream of the Wadi El-Arab Dam, tap the B2/A7 aquifer. They discharge artesian, thermal and partially mineralized water.

The main recharge of the aquifer takes place in the highlands of Irbid and Ajlun and in the Jabal Arab-Druz area, further to the north-east beyond Jordan's territories. The deeper lying aquifers of A4 and Kurnub leak some water upwards into the B2/A7 through their overlying aquicludes because their groundwater pressure heads are higher than that of the B2/A7 (El-Nasser 1991). El-Nasser (1991) calculated a recharge to this aquifer of 127 MCM/year, with base and spring flows of 100 MCM/year. This calculation includes the groundwater resources in the areas of Wadi Yabis, Wadi Jurum, Wadi El-Arab and the Yarmouk River.

Extractions of 73 MCM/year from the B2/A7 aquifer indicate that the aquifer is being over-pumped by around 56 MCM/year. The National Water Master Plan of Jordan (NWMP 1977) calculated the renewable groundwater amount which does not appear as base flow to be 23 MCM/year. The Water Authority of Jordan gives an estimate of 47 MCM/year of available groundwater in this area.

The unconfined portion of the aquifer contains groundwater suitable for different uses. In the confined portion as a result of higher temperatures and pressures dissolution of minerals adds more salts to the groundwater. Pumping or upward leakages from the aquifer to the ground surface releases the pressure and causes disequilibrium in the water composition; purification is therefore required to make it fit for domestic uses.

Uranium minerals in the Muwaqqar Oil Shale and in the Silicified Limestone Formations together with the daughter elements produce high concentrations of radon gas that are dissolved in the water, making water treatment a necessary condition for domestic use. Lime treatment in the Wadi Arab water purification plant seems to take care of the other uranium daughter elements by precipitation.

Mukheiba and Wadi El-Arab wells produce water which requires chemical treatment, because it is in chemical disequilibrium precipitating carbonates into the supply system and water use facilities. The water network of the Irbid governorate supplied from that water has suffered from this shortcoming since 1986.

Analyses performed on the water indicate that the water quality in the unconfined portions of the aquifer (Nuayma and Yarmouk University wells) means that it is suitable for all uses, although some signs of pollution have become more evident as indicated by increasing concentrations of nitrates and phosphates in the water.

Water treatment to rid drinking water of the high radon concentrations in Mukheiba and Wadi El-Arab wells is mandatory because this is a carcinogenic agent.

2.2.4.2   Southern Part of the Jordan Valley Escarpment

Upper Cretaceous limestone and Lower Cretaceous sandstone build the aquifer in this area. The area is desiccated into blocks due to its recent geologic history and the formation of the Jordan Rift Valley. These blocks may extend for a few kilometers in length and width causing the groundwater table in the limestone aquifer to be discontinuous, and the groundwater is mainly found in blocks separated from each other by faults or other discontinuities. The general groundwater flow is directed from the highlands towards the Jordan Valley where it discharges, either through springs or seepages, or by flowing laterally into the recent deposits occupying the Jordan Valley floor.

The aquifers are recharged by precipitation along the highlands of Amman and Balqa (Salt). The water balance shows that the amount of available, renewable groundwater in excess of base flow can be estimated at 10 MCM/year.

The groundwater in the deep aquifer formed by the Lower Cretaceous sandstone is confined and mineralized, as has been shown by wells drilled in the Kafrain, Rama and Wadi Hisban areas which produce artesian water with salinities of up to a few thousand mg/L.

The groundwater in this aquifer flows laterally into the recent sediments of the Jordan Valley area and leaks upwards through them to the earth surface and discharges along the lower reaches of the Jordan River side wadis.

The flowing wells drilled into this aquifer, which produce mineralized thermal water, are not used and their discharges are not controlled. Therefore, they are draining and depleting the aquifer which will certainly result in depriving the overlying Upper Cretaceous aquifers of their water support system and hence lead to a continuous drop in their groundwater levels and to depletion of resources.

The unused water discharges of these flowing wells drilled at the foothills of the Jordan Valley escarpment should be stopped as a conservation measure of the groundwater resources of the country.

The water of the Upper Cretaceous aquifers is of good quality and is suitable for the different common uses. But, since the catchment area is continuously becoming more urbanized, some signs of groundwater pollution are becoming evident (Hisban and Wadi Sir springs).

The groundwater of the deep aquifer is mineralized, thermal and artesian. It discharges CO_2, H_2S and radon gases in addition to its high iron contents which form a scale upon the water coming into contact with the atmospheric air.

In its natural state the groundwater of this aquifer can only be used to irrigate salt-tolerant crops. For any other use it should either be mixed with less saline water or be desalinated.

2.2.4.3 Jordan Valley Floor Area

Recent sediments of the Neogene and Quaternary periods cover the floor of the Jordan Rift Valley. These sediments are generally unconsolidated and possess high porosity and permeability, thus building extensive aquifers. These friable sediments inter-finger with salty, clayey deposits such as the Lisan Marls deposited in the ancestor seas of the Dead Sea, which extended northwards and southwards tens of km beyond the present shores of the Dead Sea (Horowitz 1979). The groundwater flow along the foothills of the eastern mountains is directed towards the Jordan River course (Fig. 2.27).

The amount of available groundwater in this area ranges from 18–20 MCM/year (WAJ 1991; NWMP 1977). The water quality in the northern Jordan Valley area is generally good and suitable for irrigation; in certain parts (Suleikhat to Yabis) it is

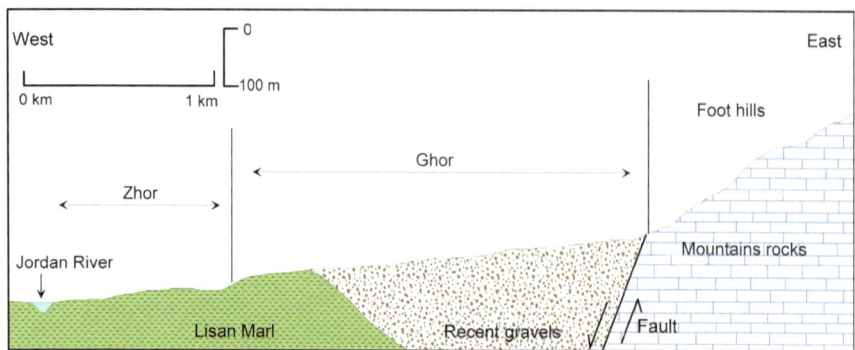

Fig. 2.27 Simplified geologic E-W cross-section showing the general relationship between the highland aquifers and the recent deposits in the Jordan Valley area (Middle part of the Lower Jordan Valley)

even suitable for drinking. In the southern part of the lower Jordan Valley, south of Zaqra River, the salinity of the groundwater increases due to the presence of saline formations and due to irrigation return flows. In the western parts, east of the Jordan River course, the water salinity goes up to a few ten thousand parts per million (Wadi Mallaha). Locally, the alluvial fans at the mountain foothills contain water of excellent quality that can be used for all purposes (Figs. 2.28 and 2.29).

2.2.4.4 Amman Zarqa Area

The main aquifer in this area consists of the Amman-Wadi Sir B2/A7 limestone and chert aquifer and because it crops out at ground surface it is partly eroded and covered by recent wadi sediments building one aquifer system. Separated from the B2/A7 by the underlying Shueib aquiclude Formation A5, 6 the Hummar Formation A4 builds the deep aquifer in the area.

Amman Zarqa groundwater basin can be divided into two parts; an eastern part extending to the north-east of Zarqa River and a western part extending to the west of it. This division is important because of the different groundwater flow systems prevailing in the area. Zarqa River forms the effluent stream of the area's groundwater. Groundwater in the eastern part flows in a westerly direction and that in the western highlands of Amman and its surroundings flows in an easterly direction. Both groundwater flows converge in the Zarqa River area and discharge in the form of springs or they flow underground along Zarqa River.

The renewable groundwater amounts in the basin average 88 MCM/year; 35 MCM/year issue at the ground surface as spring water along the Zarqa River course, and the remaining 53 MCM/year are extracted from wells distributed over the basin.

At present the groundwater in the basin is over-exploited, especially in its eastern parts, such as the subareas of Dhuleil and Khalidiya.

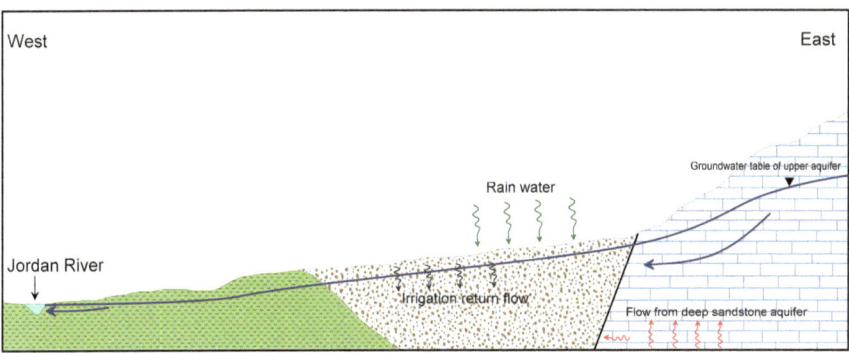

Fig. 2.28   Recharge sources to the alluvial aquifers in the Jordan Valley area

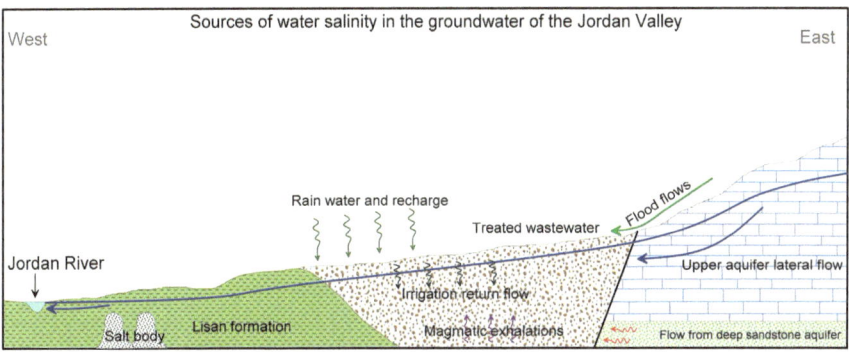

Fig. 2.29   Sources of water salinity in the alluvial aquifers in the Jordan Valley area

Recharge to the eastern parts of the aquifer comes from precipitation falling over the area and from flood waters flowing within the area. Irrigation activities, especially in the Dhuleil area, result in irrigation return flows which infiltrate back to the groundwater body causing its pollution.

Recharge in the western part of the aquifer originates from precipitation over the highlands of Amman and its surroundings, flood flows along wadi courses, leakages from the water supply system and return flows from household and commercial uses.

Irrigation, and industrial and domestic return flows contribute to the groundwater at about 40 MCM/year.

In the Amman-Rusaifa area about 30% of the discharged groundwater consists of recycled water, returning from leaky water supply systems, cesspools, sewerage systems and other uncontrolled return flows.

In the Dhuleil area, irrigation return flows during the last four decades, of several million cubic meters per year, have percolated down to join the groundwater gradually leading to its salinization. Downstream of Khirbet-es-Samra, along the course of Wadi Dhuleil the groundwater levels have risen by more than 25 m during the last three decades as a result of infiltrating treated and untreated waste waters from

Khirbet-es-Samra waste water treatment plant and the groundwater there has become unsuitable for almost all uses including irrigation.

The groundwater qualities in the area are affected by a variety of factors such as: recharge and discharge mechanisms, inflows of treated and untreated waste waters, mixing of different water qualities, leaching of solid wastes and others. Therefore, the groundwater quality of the basin differs from one area to another and no general water quality can be assigned for the whole catchment. But, large and major subareas can be delineated by the features that characterize each of them (see chapter on pollution).

2.2.4.5   Dead Sea Area

This area extends east of the Dead Sea to some 50 km into the plateau. Two aquifer complexes underlie the area, namely: the upper limestone aquifer complex and the lower sandstone aquifer complex.

The Upper Aquifer (Calcareous Aquifer)
The upper aquifer crops out along the high mountains where it receives precipitation water which infiltrates through the soil and rock covers to the groundwater and discharges in short time periods, measured in years. The total available renewable groundwater amounts to around 87 MCM/year. Almost half of it discharges as spring water along the upper reaches of the Dead Sea wadis such as Zarqa Ma'in, Wala, Mujib, Karak, Shgeig and Ibn Hamad. Groundwater is also pumped from the aquifer through wells in the highlands in the areas of Madaba, Mujib, Qatraneh and Karak for domestic use and irrigation purposes.

The Lower Aquifer (Sandstone Complex)
This aquifer does not receive major direct recharge by the infiltration of precipitation along water courses. The water in it originates from lateral and downward leakages from other areas and aquifers. Because the sandstone aquifer complex (which extends under the entire area of Jordan) crops out along the deep-lying Dead Sea wadis and shores, the water in the different aquifers tends to flow to the Dead Sea area, the ultimate base level for the entire Levant area.

The total discharge of the lower aquifer sandstone complex at the eastern slopes to the Dead Sea was calculated by Salameh and Udluft (1985) to be around 90 MCM/year of mostly thermal mineralized water (Fig. 2.23).

The groundwater is generally thermal, contains some trace metals and has elevated salinity, thus allowing only restricted use such as the irrigation of salt-semi-tolerant crops. This thermal water is suitable for therapeutic purposes; together, the prevailing climate in the Dead Sea area and the thermal spring water issuing from the sandstone aquifer complex represent a potential source of wealth for the country if appropriately developed for these purposes.

The water quality of the upper aquifer differs widely from one area to another. The concentrations of the different parameters are relatively small and within the acceptable standards for the different uses in the recharge areas increasing rapidly along the groundwater flow directions and along some geologic structures because of mixing processes with upward-leaking lower aquifer groundwater.

Pollutants originating from irrigation return flows and waste water seepages have not yet affected the aquifer, but some signs indicate their effects such as slight increases in the nitrate and phosphate levels.

The groundwater discharged from both the lower and the upper aquifers estimated at 45 MCM/year is captured at the entrance of the different wadis into the Dead Sea area, such as Mujib and Zarqa Ma'in, and pumped for treatment so that it can be added to the municipal water supply.

2.2.4.6 Northern Wadi Araba

Wadi Araba floor is covered by alluvial sediments brought in from the surrounding mountains in the east and west with thicknesses of thousands of meters. The water in the recent alluvial sediments is fresh to brackish but at greater depths it is saline due to the effects of the geologic history of the area, especially the former extensions of the Dead Sea ancestors to about 45 km to the south of the present Dead Sea southern shore (Fig. 2.30).

In detail, the groundwater here is found in the alluvial deposits, and in talus and alluvial fans with a total thickness of about 250 m. The groundwater is recharged along the eastern highlands and flows from there in a westerly direction. In northern Wadi Araba it then flows in a northerly direction towards the Dead Sea.

The throughput of groundwater from this area into the Dead Sea was calculated at around 22 MCM/year (Abu Zirr 1989). Generally, the water salinity increases in the direction of groundwater flow; from central Wadi Araba to the Dead Sea. In addition, irrigation return flows, especially in the Feifa and Safi areas, are gradually leading to groundwater quality deterioration as indicated by increasing groundwater salinity and higher phosphate and nitrate contents.

2.2.4.7 Southern Wadi Araba

Southern Wadi Araba floor is covered by alluvial sediments brought in from the surrounding mountains in the east and west with thicknesses of thousands of meters, but the fresh to brackish groundwater is present in the uppermost near-surface parts of the aquifer.

The groundwater here originates from infiltrating precipitation water falling over the eastern highlands. From there it flows in a westerly direction towards Wadi Araba and joins the fluviatile and alluvial deposits covering the Wadi Araba floor by

Fig. 2.30 Groundwater
flow system in Wadi Araba

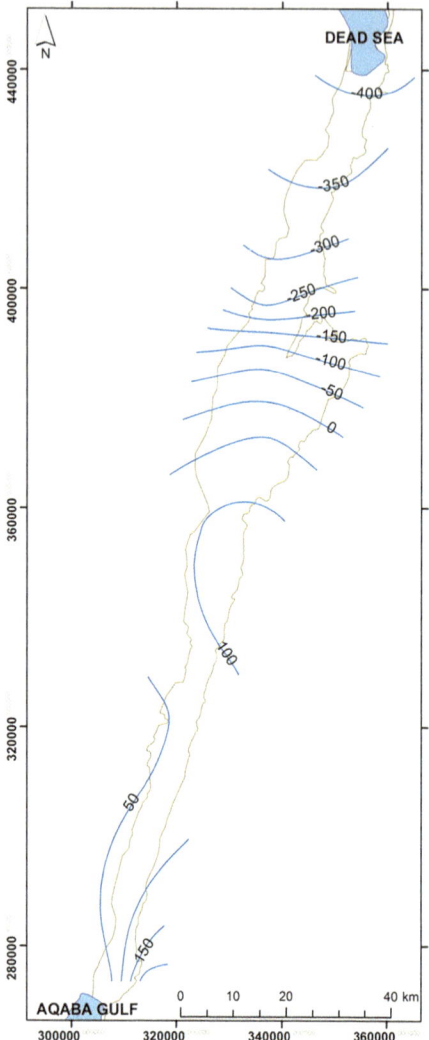

laterally flowing into them and from there it then flows southwards to the Red Sea.
Only a very small part of the recharge takes place in Wadi Araba itself due to limited
precipitation amounts.

The aquifer throughput into the Gulf of Aqaba is calculated to be around
10 MCM/year.

The groundwater quality resembles that of the northern part of Wadi Araba. The
salinity of the groundwater increases in the direction of flow; from north to south.
In southern Wadi Araba only limited agricultural activities have been developed.
The climate of the area is very harsh with summer temperatures reaching 50 °C and
relative humidity of 10–15%, making life very unpleasant.

2.2.4.8   Disi-Mudawwara Area

The Disi-Mudawwara aquifer system consists of sandstone and shale. It crops out in the Disi-Mudawwara area and underlies the entire area of Jordan at different depths overlying the basement complex with increasing thickness in northerly and north-easterly directions. It also extends southwards into Saudi Arabia. The basement complex consists of intrusive igneous rocks functioning as an aquiclude. The Disi aquifer is a medium- to fine- grained sandstone measuring about 1000 m in thickness. The average precipitation over the area is around 80 mm/year, with an average potential evaporation of 4000 mm/year.

The groundwater in the western parts of the Disi-Mudawwara area is unconfined and its level is found at depths of around 80 m below ground surface, whereas in the eastern Mudawwara area, the water is confined and used to be partly artesian. The groundwater flow is directed towards the north and north-east (Fig. 2.20). The average permeability of the aquifer is 2.23×10^{-5} m/s with a groundwater gradient of 0.143%. When assuming a groundwater flow of 50 km in width and a saturation depth of 1000 m, the throughput of the aquifer can be calculated to be: $1.68 \times 10^{-5} \times 1.43 \times 10^{-3} \times 50 \times 10^3 \times 10^3 \times 30.5 \times 10^6$ or about 50 MCM/year. This figure could give an indication about the average recharge of the aquifer.

C^{14} age determination of samples collected from that area gave a range of 11,000–13,000 years. The groundwater in the Disi-Mudawwara area receives only limited recharge, but it seems to be adequate to keep the groundwater flowing. The aquifer contains several billion cubic meters of water, but most of this is fossil water and extraction from it comes at the expense of storage; in other words, extraction from it is a mining process.

Naturally, the groundwater in Disi has a very low salinity and is free of pollution signs of any type, because the area is not industrialized and has a very low population density. Irrigation projects in the area are expected to cause irrigation return flows to infiltrate to the groundwater which will result in salinity increases and to pollution by agricultural pollutants such as fertilizers.

Generally, aquifers composed of disintegrated granitic rocks contain radioactive substances in the form of uranium and disintegration daughters such as radon and radium. Therefore, the water in them also contains radioactive substances. When this water is to be used for drinking, treatment may be required for water from certain wells, but not for all of them. Mixing with other waters with low radioactivity is also a very viable solution.

The Disi-Mudawwara aquifer extends in the underground of Jordan. It underlies all geologic units, and is hydraulically interconnected with all overlying aquifers. This implies that extraction of groundwater from Disi-Mudawwara aquifer at any location in Jordan, except the Disi Mudawwara area will result in the lowering of groundwater levels in overlying aquifers.

Extraction of water in the Disi-Mudawwara area increased from 15 MCM/year in 1983 to about 100 MCM/year after 1995 until 2013, when 100 MCM/year started to be pumped to satisfy drinking water needs in the northern parts of the country and when water used for irrigation in the area was reduced by curtailing farming activities. Extractions at present are estimated at 150 MCM/year; around 130 MCM/year for the municipal supply of Aqaba and north Jordan, and 20 MCM/year for irrigation and recreation purposes. The total amount of extraction starting from the 1980s can be estimated at around 3100 MCM, causing a non-recoverable drawdown in water levels ranging from 10 to 40 m. This non-recoverable, irreversible decline is a major warning concerning the continuity of water resources.

The final results of the hydrological studies on the Disi-Mudawwara aquifer can be summarized as follows:

- The groundwater is fossil water and its recharge is estimated at 50 MCM/year. The age of the groundwater has been measured to be over 10,000 years.
- The total pumping of about 3100 MCM over the last few years has caused a drop in water level amounting to more than 20 m on average.
- The groundwater body of Disi-Mudawwara underlies the entire area of Jordan and is, on a regional scale, hydraulically interconnected with all overlying aquifers. Hence, extraction from this water body will affect all the groundwater resources in Jordan, as it lies in the up-gradient direction, and will ultimately, in tens of years, cause a general drop in the water levels of the country.
- There are indications that the water salinity has started to increase due to salt releases from the overlying Khreim confining unit.
- The Disi-Mudawwara aquifer is the only strategic water reserve of Jordan.

2.2.4.9 The Azraq Area

The Azraq area forms the northern extension of the NW-SE trending geological and geomorphologic Sirhan depression. The Azraq area functions as a base level for both surface and groundwater of its wider surroundings which used to collect there to form an oasis.

The groundwater in this area is found in different aquifer systems ranging from recent deposits to the deep sandstone aquifer complex. The shallow aquifer of recent deposits; basalt deposites and parts of Rijam and Shallala (B4,5) aquifers contains renewable water. The recharge area of this aquifer lies mainly in the area covered by basaltic rocks in the northern and north-western part of the basin. The recharge to the intermediate aquifer, which consists of the Upper Cretaceous B2 and A7 Formations, lies also in the basalt area of Jabel Druz and in the highlands of Amman-Madaba-Karak-Tafilah. Therefore, it takes a long time for the groundwater to reach the Azraq area and hence its old age; thousands of years.

The groundwater in the deep sandstone aquifer complex has a relatively old age because it takes a long time to cross the distance from the source areas in the south to Azraq area which can be estimated at a few thousands of years. The source areas of the groundwater are:

- The highlands of Amman – Madaba-Karak and Tafilah, where precipitation water infiltrates into the Upper Cretaceous rocks outcropping there, flows in an

easterly direction and percolates down to the lower sandstone aquifer complex recharging its groundwater.
– The highlands of Tafila and Shoubak, where precipitation water infiltrates into the exposed rocks consisting mainly of Upper Cretaceous sediments and flows underground across the Jafr basin towards Azraq.
– The Jabal Arab-Druz area where precipitation water infiltrates and percolates down into the deep sandstone aquifer, and flows towards the Azraq depression area.

These groundwater components join in the underground of Azraq and flow in the sandstone aquifer complex in a westerly direction towards the Dead Sea to discharge along the Dead Sea eastern slopes as mineralized thermal groundwater (compare area 5).

The groundwater in the different aquifers that overlie each other, from the shallow aquifer to the deep sandstone complex, is a hydraulic connection moving in the following directions:

– The groundwater flow in the shallow aquifer is directed towards Azraq Oasis with only a lateral component.
– The groundwater flow in the intermediate aquifer flows from west, north and south towards the Azraq Oasis. Part of that groundwater leaks upwards into the shallow aquifer, and another part leaks down into the sandstone aquifer complex. A third part flows further eastwards to Saudi Arabia.
– The downward leakage of the overlying aquifers into the deep sandstone aquifer complex, as well as the groundwater in that complex which flows in laterally coming from the south, north and eventually the east, then flows in westerly and eventually north-westerly directions to reach the Dead Sea and probably the Jordan Valley areas.

The amount of groundwater available in the shallow aquifer is calculated to be 20–24 MCM/year.

Due to over-pumping, the water levels in the surroundings of the oasis have dropped by a few tens of meters which has resulted in the drying of the oasis, in the cessation of the discharges of Qaisiya, Soda and other springs which used to feed the oasis and in increasing groundwater salinities.

The groundwater of the shallow basalt aquifer is of a very good quality for different uses. The intermediate aquifer water quality depends on the depth and site within the aquifer. A well drilled into the intermediate aquifer in the Azraq area (AZ1) produced artesian water with a salinity of 1500 μS/cm. The well's water contained carbon dioxide, hydrogen sulphide and radon gases, indicating the confined nature of the aquifer in that area, leaking upward and downward to the overlying shallow and underlying deep aquifers.

The upper shallow aquifer in Azraq is now being heavily over-exploited with some 600 private and governmental wells resulting in dropping water levels by tens of meters and rapidly rising salinities.

2.2.4.10   Jafr Area

Rijam (B4) Formation of the Balqa Group, consisting of thin beds of chert, limestone, clay and marl with a total thickness of 20–25 m forms the main groundwater
aquifer in the area.

The deeper aquifers B2/A7 and the Kurnub and Disi sandstones underlie the area
and are separated from each other by thick aquitards. The three aquifers are hydraulically weakly interconnected. The groundwater flow in the B4 aquifer is generally
directed from west and south-west to the center of the Jafr depression (Fig. 2.31). In
the deep aquifers, the groundwater flow is directed in a general northerly direction
with flow components towards the north-east, north and north-west. The groundwater in the deeper aquifers builds the support and backbone of overlying groundwater
bodies found further north-west and south of Jafr basin. Therefore, extracting
groundwater from the deep aquifers would undermine overlying aquifers' groundwater and is in its final result quasi an extraction from the shallow B4 aquifer.

The B4 aquifer receives recharge in the highlands of Shoubak lying west of Jafr
basin. Direct recharge by precipitation in the Jafr depression area is negligible
because the surface area of the playa is covered by very fine sediments which do not
allow infiltration and groundwater recharge.

Parker (1970) estimated the total recharge to the B4 aquifer at around 7 MCM/
year. In the late 1960s the over-exploitation of the groundwater resources led, after
only a few years of extraction, to deterioration in the water quality. The salinity

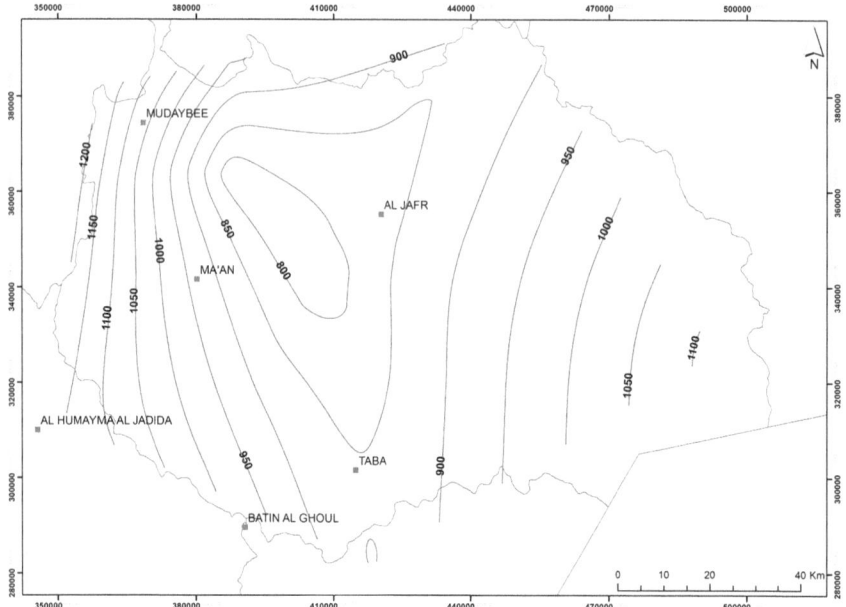

Fig. 2.31 Groundwater flow pattern in Jafr basin

increased rapidly from 600–700 mg/L in the early 1960 to values between 700 and 2800 mg/L in the early 1970s.

After that time, the groundwater quality stabilized due to limited extraction from wells, although some water salinities increased beyond the level at which they could be used for irrigation.

Jafr basin was the first main groundwater area in Jordan to suffer from over-exploitation of groundwater as expressed in groundwater resources depletion and salinization. Other groundwater basins such as Dhuleil, Aqib, Za'atari and Azraq were also exposed to depletion and salinization.

2.2.4.11 Sirhan and Hamad Areas

The shallow composite aquifer consisting of Upper Cretaceous and Tertiary rocks in addition to recent sediments of wadi fills, basalts and alluvial deposits build the main aquifer in these basins. The base level for the groundwater is the Sirhan depression extending in a south-easterly – north-westerly direction (Fig. 2.32).

The permeability of the aquifers of both areas is very small. Rock units extending from the highlands of Jordan towards the east show, due to their depositional environment, a general decrease in grain size and a general increase in siltation and

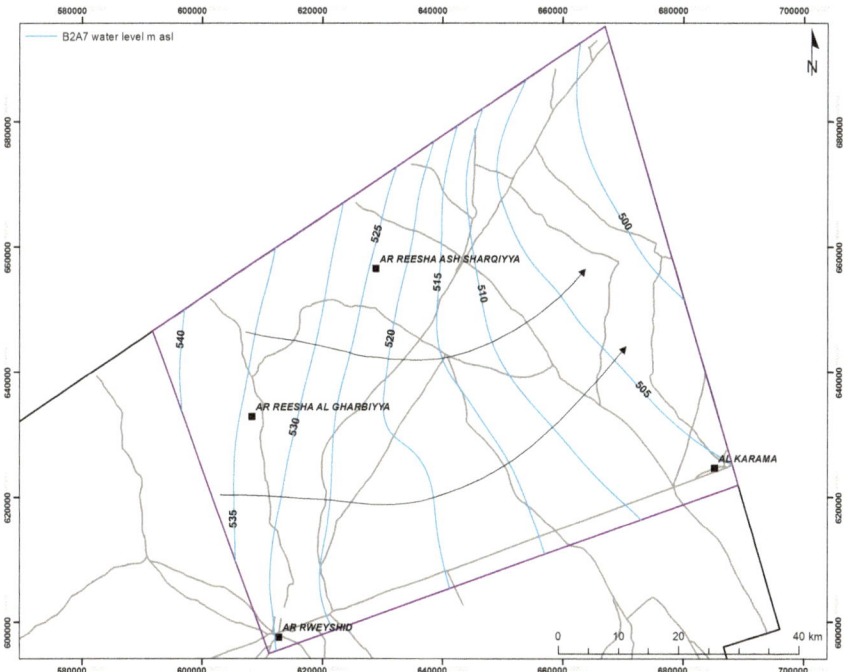

Fig. 2.32 Groundwater flow pattern in Hamad basin

cementation. The result of this is that aquifers are not well developed and the water movement through them is very slow. Therefore, wells drilled in these aquifers have small yields and the long intensive interactions of the groundwater with aquifer rocks, which are prone to dissolution, results in higher groundwater salinities.

The estimated available groundwater resources in the Sirhan and Hamad areas are 5–10 MCM/year distributed over a large area. Hence, the groundwater is considered as sparsely available and can only be used for restricted local development (ACSAD 1980). Generally, in addition to its scarcity, the groundwater in this area suffers from a salinity problem which ranges from around 1000 μS/cm up to 4500 μS/cm with the majority of sources having salinities of about 2000 μS/cm.

2.2.5 Thermal Mineralized Water

Jordan is generally a country with scarce water resources but it enjoys the presence of abundant sources of thermal and mineral waters distributed all over its territories and especially along the Rift Valley.

The physical and chemical properties of these waters and the geologic conditions governing their presence have been investigated; and a comparison with other similar sources in the world has been made to indicate their potential uses in heating, therapeutic purposes, fish farming, irrigation etc. and to illustrate the scientific and economic relevance of these sources, which may lead to their exploitation for the benefit of the local population and the country as a whole.

The sites of thermal waters in Jordan along the Rift Valley enjoy special climatic conditions with long dry summers and cool wet winters and with air oxygen concentrations increasing with decreasing elevation towards the Dead Sea (at 430 mbsl in 2016). In that same direction stronger absorption of waves takes place, before the radiation of the sun reaches the ground surface, because of the thickness of the atmospheric layer which increases in that same direction and the sun radiation that attenuates in it, especially the UV-waves, which decrease by about 12% compared with the highlands further east.

2.2.5.1 Potential Uses of Jordan's Thermal Mineralized Springs' Water

Thermal water

- Heating of buildings and greenhouses and fish production
- Therapeutic agent

Water containing hydrogen sulphide gas

- This water benefits blood circulation and the nervous system.

Gaseous water containing carbon dioxide

- water containing CO_2 is a source of mineral water that can be used for drinking and as an agent for enhancing blood circulation. It is used to Meaning unclear 'cure disturbances in blood circulation to the body and limbs.

Bromide-rich water

- This water can be used as a tranquilizing agent and helps relaxation.

Radioactive water

- This water can be used for enhancing general physical health, and for numerous rheumatic diseases and in certain cases of sterility, fertility problems and gynaecology. This type of water should only be used externally.

Iodide water (internal, drinking)

- Waters containing iodide at a concentration which allows them to be classified as iodide waters can be used to enhance blood circulation, and for respiratory tract inflammations, heart diseases and eye diseases.

Iron containing water (internal, drinking)

- Such waters are used to alleviate stomach, kidney and intestinal complaints, to improve blood circulation and to aid the digestive process.

Sodium – Chloride – Sulphate water (internal and external)

- Such water is used to treat diseases of the stomach, kidney and intestines and the nervous system. It is also used for rheumatic diseases.

Calcium – Magnesium – Sulphate water (internal inhalation and external use)

- Such water is used to treat rheumatic and heart diseases and diseases of the nervous and respiratory systems.

Sodium- Chloride water (internal inhalation and external use)

- Such water is used to treat the respiratory organs in general and the respiratory tract in particular, rheumatic diseases and diseases of the gynaecological and nervous systems.

In Jordan, there are about 200 thermal springs distributed across the country. They are concentrated along the eastern side of the Jordan Rift Valley in Zarqa Ma'in, Zara, Afra, Weeda'a, and in the north Shuna Himma areas.

The composition of these springs' waters differs from one area to the other and accordingly so do their therapeutic agents such as heat, H_2S, radon, iron, iodide, bromide, calcium, magnesium, sulphate, sodium etc. and combinations of these therapeutic agents (see Table 2.2).

Table 2.2   Composition of the main thermal waters in Jordan

Variable	Al-Shallal	Al-Amir	Zarqa River	Hisban	N. Shuna	Al-Maqla	Afra
EH-value	15.8	18.9	22.8	19.4	12.3	11.45	nm
Temp °C	56.6	48.6	47.1	31.8	52.7	41.4	46.6
pH-value	6.3	6.2	6.76	6.37	7.06	7.12	7.12
EC µS/cm	3051	3080	1473	4190	981	1336	563
TDS mg/L	2279	2346	1114	3840	863	1117	297
Na^+ meq/L	19.52	18.63	7.79	29.01	3.3	5.86	36.41
K^+ meq/L	1.11	1.32	0.61	2.42	0.12	0.43	2.16
Mg^{+2} meq/L	3.22	3.41	2.05	8.82	3.66	2.69	17.91
Ca^{+2} meq/L	7.23	7.82	4.93	11.84	4.07	6.24	47.18
Cl^-meq/L	21.52	22.32	9.17	28.04	2.83	6.03	69.38
NO_3^- meq/L	0.1	0.05	0.04	0.04	0.02	0.05	0
SO_4^{-2} meq/L	3.82	3.93	2.51	7.7	1.75	3.44	66.17
HCO_3^- meq/L	4.8	4.82	3.12	13.1	6.37	5.53	133.1
CO_3^{-2} mg/L	215	224	50	416	59	79	102
F^- mg/L	0.31	0.43	0.22	0.61	0.61	0.81	0.2
Br- mg/L	7.74	7.21	4.02	6.44	0.91	3.13	0.03
I^-mg/L	0.11	0.08	0.08	0.14	0.5	0.1	0.005
TC meq/L	31.1	31.2	15.4	nm	11	15.2	nm
Fe mg/L	0.0.09	0.23	0.11	1.34	0.12	0.18	0.1
Mn mg/L	0.6	0.79	0.56	0.13	0.007	0.008	0.003
Cd mg/L	0.01	0.005	0.003	0.0006	0.0024	0.0027	nm
Zn mg/L	0.06	0.049	0.024	0.015	0.002	0.018	0.002
Pb mg/L	0.02	0.023	0.13	0.027	0.03	0.03	0.002
Rn nCi/L	4.1	20.6	25.5	15.8	10.2	31.5	19.4
NH_4^+ mg/L	0.9	1.14	0.56	4.21	1.15	2.62	nm
H_2S mg/L	0.2	0.16	0.04		5.8	9.5	0.3
Br mg/L	7.2	7.2	3.9	6.4	0.91	3.3	0.75

nm not measured

References

Abu-Zirr M (1989) Hydrochemistry and hydrogeology of the central part of Wadi Araba. M.Sc. thesis, University of Jordan, Amman

ACSAD (1980) The Arab centre for the studies of arid zones and dry lands, water resources of the Hamad basin, Damascus (Arabic)

Alhejoj I, Salameh E, Bandel K (2014) Mayflies (Order Ephemeroptera): an effective indicator of water bodies conditions in Jordan. Int J Sci Res Environ Sci 2(10):361

Alhejoj I, Bandel K, Salameh E (2015) Floral species as environmental quality indicators in Jordan: high salinity and alkalinity environments. J Environ Prot 6(05):494

Bandel K, Salameh E (1981) Hydrochemical and hydrobiological research of the pollution of the waters of the Amman Zerka area (Jordan). Deutsche Gesellschaft für Technische Zusammenarbeit

BGR and WAJ (1995) Groundwater resources of Northern Jordan. Vol. 2, Part 2: monitoring of groundwater levels in Jordan. Water Authority of Jordan (WAJ) and Federal Institute for Geosciences and Natural Resources (BGR), BGR-Archive No. 112708, Amman, Jordan

El-Nasser H (1991) Groundwater resources of the deep aquifer system in NW Jordan – Hydrogeological and hydrochemical quasi 3-dimensional modelling, Hydrogeology and Umwelt, H.3, Wùrzburg, Germany

Horowitz A (1979) The quaternary of Israel. Academic, London

MWI (Ministry of Water and Irrigation) Jordan. Open files (2016)

NWMP (1977) National Water Master Plan of Jordan, vol 8. Bundesanstalt für Geowissenschaften und Rohstoffe/GTZ, Hannover/Eschborn

Parker DH (1970) Investigation of the sandstone aquifers of East Jordan. UNDP, AGL: SF/Jor. 9, Rome 1970

Salameh E (1996) Water quality degradation in Jordan. Friedrich Ebert Stiftung, Amman and Royal Society for the Conservation of Nature, Amman, 179 p

Salameh E, Bannayan H (1993) Water resources of Jordan – present status and future potentials. Friedrich Ebert Stiftung, Amman, 183 p

Salameh E, Udluft P (1985) The hydrodynamic pattern of Central Jordan. Geol Jb 38:39–53, Hannover

Water Authority of Jordan (WAJ). Open files of the Water Authority of Jordan/Amman (1991)

Chapter 3
Patterns of Water Use

3.1 Water Use

3.1.1 Domestic Uses

The present total water supplied for household use in Jordan is 420 MCM/year. This includes the physical losses from the water supply network incurred by corrosion and damage and the stealing of water (water theft). According to the Ministry of Water and Irrigation, these total losses (unaccounted for water) amount to 47% of all the supplied water (MWI 2016). Most of the stolen water is used in irrigation and only a very small percentage for household uses.

The average per capita household water use in Jordan is calculated to be around 80 L/day taking into consideration physical losses, illegally used water and commercial supplies from the household supply network of 20 MCM/year. Compared to the household uses in Europe of 130–250 L/c.d., to those of Israel of 280–300 L/c.d., to the Gulf States of 280–450 L/c.d., and to Iraq, Syria and Egypt of 130 L/c.d., it can be concluded that Jordanians' usage is low. This is not only because they are extremely concerned about water, but also because more water is simply not available. During the summer, when household water use is higher than in winter and water supplies are limited, 85% of Jordanians live at the hygienic brink in their use of water and less water would be detrimental to public health.

Guest workers in Jordan work mainly in the agriculture sector or in the construction industry and therefore most of their water use is taken from the water supplies of these sectors.

Refugees in Jordan, mainly Syrians, are supplied with water through sources especially developed for them. In general they use less water than Jordanians because of their present living conditions in refugee camps where there are no recreational or gardening activities, no luxuries and no losses, because of the new supply network and absence of illegal connections. Their per capita use is estimated at 40–50 L/c.d. The development of water uses in the time period 200–2013 is shown in Fig. 3.1.

The text of this chapter is based on Salameh and Banayan (1993) and Salameh (1996). Measurements and numbers are updated (MWI open files and from own analyses).

© Springer International Publishing AG, part of Springer Nature 2018   61
E. Salameh et al., *Water Resources of Jordan*, World Water Resources 1,
https://doi.org/10.1007/978-3-319-77748-1_3

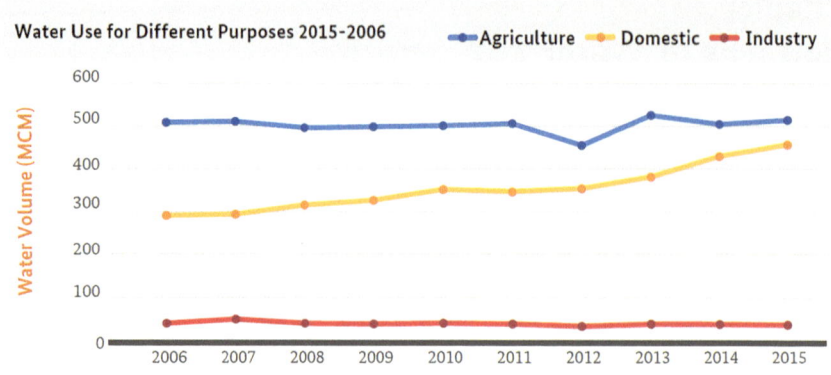

Fig. 3.1 Development of water use in the different sectors 2006–2015

3.1.2 Industrial Uses

Around 45 MCM/year of water are presently used for industry in Jordan, of which
37 MCM/year are extracted from wells and 8 MCM/year are provided from the
drinking water supply networks. Mining industries consume the major part of this
water. During the 1980s and 1990s all industries suffered from the shortage of water
in the country, leading them to raise their water use efficiency and to recycle their
waste water whenever possible. Water availability has been throughout the last few
decades a limiting factor for industrial expansion and the construction of some
water-consuming industries, especially the mining of oil shale deposits.

3.1.3 Agricultural Uses

Irrigated agriculture is an important factor in the economy of Jordan. At present it
consumes an average of 550 MCM/year. It is a main pillar in the socio-economics
of the Jordanian population, because of its role in the history of people in this part
of the world dating back to the early development of agriculture some 9000 years
ago. The lifestyles of people have, throughout human history, been determined by
the production of food depending on the availability of water.

Irrigated agriculture is the main consumer of water in Jordan. Uses in animal
husbandry and fish farming amount to some 7 MCM/year only.

The consumption an irrigated agriculture depends on the availability of resources
such as water stored in dams, spring yields, treated waste water and others. At present around 300 MCM/year of groundwater are used for agricultural purposes, con-

Table 3.1 Water use by sector (MCM/year)

Use sector	Present uses
Domestic	420
Industrial	45
Irrigation	550
Population	6.5 million Jordanians and estimated other 2.2 million refugees and guest workers

sisting of fossil, non-renewable water and recycled and base flow water. Governmental and private sector agricultural developments have put around 61.000 hectares of land under irrigation.

Jordanian farmers are well aware of the importance of water and they have therefore introduced water-saving technologies and devices such as drip irrigation instead of the traditional furrow irrigation methods.

Other countries sharing water resources with Jordan, e.g. Syria, play a role in how much water can be used in irrigation in Jordan because such countries lie in the rivers' upstream areas and control the headwaters of rivers allowing only a certain amount of water to flow down into Jordan. Jordan's water shares in the transboundary water resources have been decreasing for about three decades which has negatively impacted irrigated agriculture in Jordan, as expressed in lower agricultural production and the lower yields of irrigated areas, with all the implications this has for the employment market and for farmers' livelihoods.

3.1.4   Total Uses

Water uses in Jordan differ from 1 year to another depending on the available resources. In 2014, the total uses amounted to 1015 MCM (Table 3.1).

3.2   Water Balance: Resources Versus Consumption

In Jordan the demand of different sectors for water cannot be met by the limited available resources. This situation is worsening over time as the difference between demand and supply is increasing. It is further exacerbated by the hundred thousands of refugees coming to Jordan from Syria, Iraq and other countries afflicted by civil wars and internal political, economic and social unrest.

Surface water resources have been developed and exploited to a large extent for use in the agricultural sector. Dams, canals and advanced irrigation systems have been constructed to enable the best use of available resources. Undeveloped water sources are very limited in quantity or restricted in quality and are also expensive to develop. New dam projects in Jordan can only calculate with a few million cubic meter capacities because wadi discharges of more than that amount of water are already dammed. This fact alone shows the situation of the country, desperate for every drop of water. The construction cost of new dams exceeds JD $10/m^3$ of water.

Nonetheless, Jordan is planning these dams in the hope of alleviating the severe water shortage in the country. The Unity Dam, planned to supply Jordan with an additional 100 MCM/year of good quality water and to moderately alleviate the water shortage problem has since its construction in 2007 only collected 10–20 MCM/year. Only by the partial abandonment of the upstream Syrian water uses during 2013–2016 were additional amounts of a few million cubic meters collected in the dam.

Water harvesting projects in the desert areas, for agricultural uses, cattle watering and some groundwater recharge have been underway for the last two decades. But, developing all the desert wadis will add only around 20 MCM/year to Jordan's available water resources.

The tapping of the country's non-renewable groundwater resources started over four decades ago and over-exploitation at present reaches around 300 MCM/year. Some groundwater basins like Jafr and Dhuleil were depleted in the 1970s and 1980s; and others like Azraq, Mujib, Yarmouk, Disi and Agib are being depleted in quantity and deteriorating in quality. The present over-exploitation If this situation continues and non-renewable groundwater is extracted at the same rate, the exhaustion of groundwater resources will certainly result within the next two to three decades.

3.2.1 Future Water Demand (Fig. 3.2)

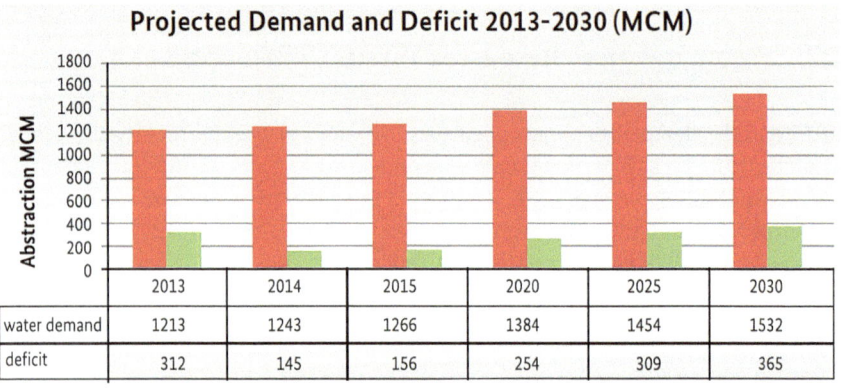

Fig. 3.2 Water use and future water demand and deficit

3.2.2   Domestic Uses

The increasing demand for domestic purposes in Jordan is caused by the following factors:

- Natural growth of population of 2.4% per year.
- Improving living standards.
- Waves of refugees settling in Jordan.

The present amount of water supplied for domestic use is around 420 MCM/ year; serving 6.5 million inhabitants in addition to 2.2 million refugees and guest workers, which is expected to rise to 9.06 million Jordanians and 3.08 million refugees and guest workers by the year 2030 (Fig. 3.2 and Table 3.2).

If supply rates per capita remain as low as they are at present and if the leaky network is not repaired and leakages stopped, 585 MCM/year will be needed by the year 2030 to cover the domestic supply of Jordanians alone.

Jordan's inhabitants consume in domestic uses an average of 80 L/c.d., which in terms of living standards and the developmental state of the country, is considered the minimum to sustain the health of the population. An average of 100–120 L/c.d. can be considered adequate to meet the population's needs. If 100 L/c.d. are consumed, and the natural population growth and the leakages continue at the same rate, Jordan will need 731 MCM/year by 2030.

In addition, improving living standards and life styles may cause a 10% increase in domestic demand.

As regards the water supply system of the country, the effects of migration and refugees' needs can only be estimated; but refugees, immigrants and foreign workers in Jordan are reflecting catastrophically on the water sector. Their water consumption at present is 64.2 MCM/year and in the year 2030, considering a growth rate of 2.4% per year, their number will be 3.06 million and their municipal water consumption will resemble that of Jordanians and will equal 90 MCM/year (Table 3.3).

A few years ago the Ministry of Water and Irrigation initiated an ambitious project to curtail as much as possible illegal water extraction and water thefts and was thus able to stop a great deal of illegal water use. If this program continues in the coming years, part of the depletion of aquifers will stop and the supply of

Table 3.2 Water use demand in million m^3 and population in million by per capita use of 80 and 100 L/c.d

Year	2016	2016	2030	2030
Consumption litters/c.d	**80**	**100**	**80**	**100**
Domestic	420	525	585	731
Industrial (self-supply)	37	37	80	80
Irrigation	550	756[a]	760[a]	860[a]
Population/Jordanians	6.5	6.5	9.06	9.06

[a]Increases are due to increasing waste water production, treatment and reuse in irrigation

Table 3.3 Present and future
water uses by refugees and
foreign workers residing in
Jordan (per capita water use
80 L/day)

Year	2016	2030
Consumption/needs in MCM/year	64.2	90
Numbers (million)	2.2	3.06

domestic water will increase as a result of curtailing water theft from the water supply system.

3.2.3   Industrial Uses

At present, industries consume around 45 MCM/year of water. The planned and expected development in this sector estimates the demand by the year 2030 to be 80 MCM/year. The main increase is expected to be caused by oil shale extraction and processing and other industries.

3.2.4   Agricultural Uses

The amounts of water used in irrigation depend on the availability of resources. This sector consumes at present some 500 MCM/year. But if new resources are not going to be developed, this amount is expected to decrease because some of the currently used water is extracted from non-renewable resources. Given the present availability of resources the agricultural sector is not expected to be allocated more fresh water and any additional amounts will have to come from treated once-used water. By 2030 agricultural water is expected to sum up to 585 MCM/year if household uses remain at their present per capita consumption of 80 L/c.d. But if consumption increases to 100 L/c.d. then more treated waste water will be available for irrigation and irrigation water will increase to 731 MCM/year.

References

MWI (Ministry of Water and Irrigation) Jordan. Open files (2016)
Salameh E (1996) Water quality degradation in Jordan. Friedrich Ebert Stiftung, Amman and Royal Society for the Conservation of Nature, Amman, 179 p
Salameh E, Bannayan H (1993) Water resources of Jordan – present status and future potentials. Friedrich Ebert Stiftung, Amman, 183 p

Chapter 4
Water Pollution

Two issues should be clearly and strictly separated from each other when considering the quality of a water source.

1. Natural water quality: water characteristics as affected by natural conditions.
2. Water quality as affected by human activities: changes in the original water qualities due to the human activities of urbanization, industrialization, agriculture and the building of structures.

For any type of water use the water quality issue is as vital as the water quantity or availability itself, with the following facts to be always kept in mind:

- Any water source can have only a limited number of uses according to its quality.
- Most water resources can serve as sources of raw water, which can be treated and manufactured to serve uses demanding certain criteria of physical, chemical or biological characteristics.
- Any water, whatever quality it has, can be treated to produce water with certain characteristics required to serve specific objectives, but the cross-cutting problem in this case is the cost of treatment.

The principle objective is whenever and wherever possible to use water sources without putting much effort or expense into treatment to reach the desired quality for a certain use. Water treatment and purification are costly undertakings and involve in many cases complicated technical processes, especially if the source undergoes continuous quality changes. Matters of water quality are highly affected by economic and financial factors. The questions are generally: how much money can we afford to spend to make a water source usable for a certain purpose; what other alternative resources are available, and is it worth investing in that specific treatment? Such questions are valid for both the original and human-affected water qualities and their treatability.

The text of this chapter is based on Salameh (1996). Analytical results and numbers have been updated.

© Springer International Publishing AG, part of Springer Nature 2018   67
E. Salameh et al., *Water Resources of Jordan*, World Water Resources 1,
https://doi.org/10.1007/978-3-319-77748-1_4

The economic equations which describe and address pollution and changes in the original water quality of a source making it unsuitable for its original uses are more difficult to judge, because its reuse and impacts have social, environmental and health aspects; issues which cannot be easily quantified.

4.1 Natural Water Qualities

Naturally, some water sources cannot be used for certain or all common uses (household, irrigation, or industrial), because their quality is impaired by one or more chemical constituent, biological content or physical property. This fact leads us to characterize the different water qualities at the source, under natural conditions (unaffected by human activities) according to their suitability for the common uses. In the following the natural water qualities in Jordan are dealt with in detail, especially because a source's quality can be a limiting factor for its use, which further limits the availability of usable resources.

4.1.1 Precipitation

In the Rift Valley area the amounts of precipitation range from 400 mm/year in the north Yarmouk River area, to 30 mm/year in the south, Aqaba. The EC of the water ranges from 40 to 500 µS/cm and the pH from 6.6 to 8.2. The pH values decrease with increasing precipitation amounts and with the progress of time in each precipitation event.

The non-availability of sufficient acid-producing gases with pH values of less than 5.5, such as NO_x and SO_2, explains the relatively high pH values of precipitation water in this part of Jordan.

In the highlands extending from Irbid in the north to Ras en Naqab in the south the annual precipitation ranges from 650 to 250 mm/year. The EC of precipitation in this area ranges from 150 to 436 µS/cm and the pH from 6.39 to 8.8. For partial sampling of precipitation events the pH ranges from 4.85 to 9, where the 5.6 value was measured at the end of very heavy rains on cold western fronts. In Amman the lead concentration of 0.021 mg/L reflects the traffic density and the leaded fuel used in Jordan.

In the eastern plateau, the precipitation rates range from 250 mm/year to 30 mm/year. The EC of precipitation water ranges from 100 to 642 µS/cm, and the pH value from 6.5 to 8.8.

Generally, precipitation water in Jordan has a pH value of more than 5.6, which is the biogenic pH value. This means that the country is not affected by any type of acid rain. The precipitation water contains high concentrations of dissolved substances as a result of dust in the atmosphere which also neutralizes any acidic, low-pH gases.

In the Rift Valley area and on the plateau both the salt contents and the pH values of precipitation are higher than those of the highlands.

This indicates that even precipitation water in Jordan as a semi-arid country contains higher salt contents which are then reflected in higher salinities of surface and groundwater resources.

Pollution in precipitation water is only found in Amman (traffic), in Zarqa (industry), in Hashimiya (refinery and thermal power station), in Fuheis (cement factory, dust), in Hasa (phosphate dust) and in Aqaba (phosphate, potash, fertilizers loading and industries).

4.1.2   Flood Flows

All over the country floodwaters, which have generally only short contact times with the country's rocks, contain small concentrations of chemical components, their physical properties; pH values, temperatures, dissolved oxygen and electric conductivities and their biological contents indicate natural conditions. Therefore, they can be used for all common uses with very minor treatment aiming at turbidity settlement and eventual chlorination as a precautionary measure in the case of household uses.

Tables 4.1 and 4.2 show the characteristics of flood flows in wadis in the different parts of the country. The floodwater salinities expressed in EC units ranged from 123 to 530 µS/cm depending on a variety of factors, such as precipitation water quality, rock and soil types and coverings.

The catchment areas, their topographic features, rainfall amounts, ground surface temperatures, and land use.

All the parameters listed in Tables 4.1 and 4.2 indicate a water quality allowing use for all common purposes – household, industry and irrigation – without any major treatment except filtration to remove the suspended load and chlorination as a precautionary measure for household uses.

4.1.3   Base Flows and Groundwater

The base flow along water courses consists of spring discharges and other seepages within the catchment area of that water course. Hence, its characteristics reflect, to a certain degree, the qualities of the groundwater underlying the drainage area.

According to their quality characteristics different groups of base flows are found in Jordan.

Group I: Base Flows and Groundwater Suitable for All Common Uses
Such groundwaters are discharged from springs all over the country generally at elevations of 500 masl and higher. The water then flows along the different wadis in

Table 4.1 Floodflow composition along wadis pouring into the Jordan Rift Valley (EC in µS/cm, NO_3^- in mg/L and all others in meq/L)

Parameter	Daba	Qastal	Zizya	Rweished	Safawi	Khalidiya	Mafraq	Muwaqqar	Azraq	Yutum	Shidiya
EC	123	212	233	229	218	291	220	186	214	135	130
pH	8.55	8.53	8.55	8.25	8.43	7.76	7.8	8.48	7.7	8.21	8.27
Ca^{+2}	1.2	1.53	1.73	1.9	1.28	1.8	0.59	1.1	1.18	0.74	1.30
Mg^+	0.4	0.69	0.72	0.4	0.19	0.26	0.45	0.2	0.20	0.13	0.35
Na^+	0.27	0.92	0.41	0.31	0.75	0.92	1.08	0.93	0.94	0.25	0.62
K^+	0.22	0.05	0.05	0.09	0.05	0.18	0.13	0.11	0.13	0.01	0.16
Cl^-	0.15	0.4	0.60	0.4	0.35	0.22	0.2	0.23	0.4	0.35	0.60
SO_4^{-2}	0.35	0.39	0.94	0.41	0.24	0.2	0.23	0.25	0.34	0.10	0.38
HCO_3^-	1.55	1.82	1.46	1.91	1.35	2.45	1.57	1.94	1.65	0.76	1.52
NO_3^-	0.54	10.2	13.8	2.1	4.2	4.8	16.2	6.8	7.2	2.4	3.2
PO_4^{-3}	2.8	0.92	0.62	0.16	0.09	0.96	0.61	0.72	0.81	0.00	0.62

Table 4.2 Floodflow composition along the plateau wadis (EC in µS/cm, NO₃ in mg/Land all others in meq/L)

Parameter	Yarmouk River	Yabis	Kufranja	Abdoun Ras El-Ain	Zarqa River Jarash Br.	Hisban	Zarqa Ma'in	Mujib	Karak	Hasa
EC	530	430	307	160	392	235	182	183	165	301
pH	7.91	8.37	8.05	8.42	8.01	7.97	8.36	7.78	7.98	8.38
Ca^{+2}	1.9	2.87	2.46	1.6	2.36	1.58	1.00	1.02	1.16	1.60
Mg^+	1.4	1.43	0.59	0.2	0.32	0.29	0.40	0.42	1.57	0.20
Na^+	1.70	0.95	1.03	0.29	1.22	0.53	0.59	0.58	1.12	1.02
K^+	0.15	0.17	0.41	0.08	0.16	0.10	0.13	0.10	0.22	0.09
Cl^-	1.58	1.10	0.4	0.25	1.20	0.50	0.23	0.27	0.78	0.39
SO_4^{-2}	0.85	0.84	0.63	0.41	0.74	0.16	0.13	0.16	1.04	2.04
HCO_3^-	2.97	2.91	2.99	1.42	2.04	1.72	1.73	1.82	2.66	2.04
NO_3^-	18.5	18.2	13.4	5.3	18.0	9.2	4.80	5.8	4.2	6.6
PO_4^{-3}	0.73	0.12	1.37	0.55	0.53	0.38	0.84	0.252	0.24	0.87

the area extending from the Yarmouk River in the north to Ras en Naqab in the south. Recharge to these springs and seepages by precipitation and floodwater takes place along the highlands extending from the Irbid area to the Ras en Naqab area with an east-west extension of 20–30 km. The recharged water discharges along the slopes overlooking the Rift Valley and its side wadis and along the wadis draining completely or partly to the east.

All springs along the higher reaches of the Yarmouk River, Wadi El-Arab, Ziqlab, Kufranja, Rajib, Yabis, Zarqa River, upstream of King Talal Dam, Shueib Dam, Kafrain Dam and Hisban diversion dam, Udheimi, Zarqa Ma'in, Shqiq, Mujib, Wala, Hasa upstream of the thermal spring discharges, Tafilah, Shoubak, Finan and Gharandal discharge a base flow with a natural quality suitable for all common uses.

Wadis and wadi systems draining completely or only their upper reaches eastward such as Zarqa, Mujib, Jafr, Wala, Shallala, and the springs of Azraq discharge, or they formerly discharged water with a quality suitable for all common uses.

Table 4.3 shows some examples of spring and base flow discharges of water suitable for all common uses. Generally, the EC is less than 1500 µS/cm; none of the other parameters has a higher concentration than that recommended by the WHO or by Jordanian standards for drinking water. Slightly elevated nitrate, calcium, chloride and phosphate concentrations indicate some type of domestic pollution.

Group 2: Water with High Salinities and/or High Temperatures
Such waters are found naturally in Jordan. According to their salinities and other properties such as temperature, pH, the presence of certain gases (carbon dioxide, radon, hydrogen sulphide) or certain elements at a high concentration, e.g., iron, manganese and others, these waters are differentiated into several subgroups:

Table 4.3 Spring and baseflow discharges along the upper reaches of wadis, unaffected or very slightly affected by domestic types of pollution (EC in μS/cm, NO_3 and PO_4 in mg/L and all others in meq/L)

Parameter	Yarmouk River	W. El-Arab	W. Ziqlab	W. Yabis	Ras El-Ain[a] Zarqa R.	ZarqaMa'in[b]	Hisban	Bahhath	Mujib	Karak	Hasa
EC	980	820	665	844	780	618	540	530	712	784	1424
pH	8.15	7.8	7.72	8.57	7.7	7.3	7.50	7.45	8.40	7.97	8.49
Ca^{+2}	3.06	5.0	2.9	2.66	5.02	3.9	3.66	2.97	3.50	1.96	3.35
Mg^+	2.56	2.6	2.6	2.56	0.75	1.7	2.08	1.29	2.10	1.68	5.71
Na^+	4.10	2.28	1.13	3.60	1.22	1.00	0.76	0.67	2.40	4.1	5.52
K^+	0.17	0.10	0.02	0.10	0.10	0.04	0.05	0.04	0.18	0.24	0.13
Cl^-	3.10	2.15	6.65	2.7	1.87	1.72	0.98	1.23	2.08	3.85	6.0
SO_4^{-2}	1.64	1.6	0.57	1.75	0.12	0.6	0.23	0.19	1.23	1.84	2.65
HCO_3^-	4.38	5.11	4.2	3.60	6.01	3.52	4.79	3.06	3.78	1.96	4.72
NO_3^-	23	21	12	11.8	20	28	24	27	15.0	13.5	11
PO_4^{-3}	0.35	0.38	0.11	0.17	0.32	0.17	0.27	0.03	0.025	0.86	0.016

[a]Ras El-Ain/Zarqa River
[b]Zarqa Ma'in in Upper Reaches

Table 4.4 Chemical analyses of waters with high salinities in the different parts of the country (EC in μS/cm, NO₃ in mg/L and all others in meq/L)

Parameter	Jordan River[a]	Wadi Mallaha	Wadi Araba UM-3	Sumaya spring Zarqa River	Omari well 2Azraq	Jafr well 17	Reasheh No. 2 Hamad	Karameh Society J. Valley
EC	8810	18,100	5300	2250	2430	5790	4330	4100
pH	8.26	7.62	6.63	7.00	8.0	7.8	7.4	7.4
Ca^{+2}	28.10	34.0	13.03	5.81	7.15	16.7	12.05	9.1
Mg^+	12.28	50.0	9.19	8.28	7.75	17.5	14.0	13.0
Na^+	42.7	117.2	16.12	9.70	9.48	23.69	20.17	17.70
K^+	2.39	5.54	0.21	0.36	0.3	0.5	0.3	0.55
Cl^-	73.75	147	29.84	14.39	12.57	46.86	19.68	31.94
SO_4^{-2}	6.94	59.6	4.255	3.08	7.85	8.10	26.67	3.40
HCO_3^-	4.02	3.96	1.0	5.02	3.75	3.32	7.13	4.20
NO_3^-	24	60	6.5	75	3.5	20	2.1	42

[a]Jordan River, King Hussein Bridge, Dry Season

Waters with High Salinities
Examples of these are the discharges of the Jordan River, Wadi Mallaha, the Azraq area at the periphery of the Azraq pools and generally, the deep groundwater in the country. These waters, according to their salinities, have only a limited range of uses. If their salinity is low enough, 1500–4000 μS/cm, they can be used to irrigate salt- semi-tolerant or tolerant crops. But, if their salinities are higher, then they cannot even be used for those purposes. Nonetheless, they can serve as a source of raw water to be desalinated or mixed with fresh water for use in different sectors. Table 4.4 shows examples of the composition of such water.

The salinity of brackish and saline water ranges from the upper limit of fresh water salinity of ca. 1500–18,000 μS/cm in Wadi Mallaha. The major constituents such as Ca, Mg, Na, Cl, and SO₄ are the main parameters contributing to salinity.

In Wadi Mallaha and Karameh Society well, both in the Jordan Valley and in Sumaya spring (Zarqa River), irrigation return flows or domestic types of pollution or both can be concluded from the higher nitrate contents.

This type of water can partly be used to irrigate salt-tolerant crops, or it can be mixed with fresh water for general irrigation purposes, or it can serve as a source of raw water for desalination.

Water with High Salinity and Elevated Temperature
This type of water results from the deep percolation of infiltration water and from very long contact times with the aquifer matrix, undergoing oxidation/reduction processes.

Such water, due to its temperature of more than 5 °C above the ambient temperature can be classified as curative water. It is generally found along the slopes overlooking the Rift Valley.

Table 4.5 Composition of thermal water with high salinity (EC in µS/cm, NO_3 and H_2S in mg/L, Rn in nCi/L and all others in meq/L)

Parameter	Zarqa Ma'in Shallala	Zara Sp. 40	Hisban thermal well	Deir-'Alla spring	Abu Thaleb spring	Abu Ziad well
EC	3050	1786	4190	3440	1928	2300
pH	6.8	6.3	6.37	6.5	6.82	6.9
Ca^{+2}	7.39	4.58	11.84	12.5	7.6	7.25
Mg^+	2.82	1.90	8.87	8.84	4.3	4.17
Na^+	14.13	7.86	29.02	13.61	7.74	10.96
K^+	1.30	0.63	2.40	0.75	0.38	0.40
Cl^-	19.07	9.26	28.05	18.45	9.30	11.71
SO_4^{-2}	4.31	2.77	7.37	12.10	4.03	4.4
HCO_3^-	3.00	2.84	13.10	6.58	7.65	6.33
NO_3^-	0.00	0.00	2.5	0.0	0.00	0.02
T °C	55.0	53.5	33	36	36.8	50.2
Rn	14.25	25.5	4.3	1.7	15.65	17.3
H_2S	0.18	0.04	Smell	Smell	Smell	0.5

Examples of this type of water are the springs of Zarqa Ma'in, Zara, Mujib, Deir-Allah and Abu Thableh and the wells of Hisban and Abu Ziad.

Table 4.5 gives the composition of such waters and some of their therapeutic agents.

Generally, these waters have salinities of more than 1500 µs/cm, a temperature of more than 33 °C. Gases such as CO_2, H_2S and radon are discharged with the water.

Such characteristics and composition allows the water to be classified as thermal-mineralized water with therapeutic properties.

This type of water can easily serve as a raw water source to be mixed with fresh water for the various uses or for desalination. Due to its high discharge temperature it could also be used for heating homes and greenhouses during the cold season.

Water with Low Salinity and High Temperatures
Thermal water discharges are not restricted to saline or brackish springs. Many springs along the slopes overlooking the Rift Valley discharge fresh thermal water. The temperature may reach more than 50 °C, but the salinity remains less than 1500 µS/cm.

Such water presents a source for all common uses after some cooling and aeration. Table 4.6 shows the composition of thermal water springs and wells with low salinities.

Radon and H_2S gases can easily be removed by aeration, which makes the water potable.

Table 4.6 Composition of thermal water with low salinities (EC in µS/cm, NO₃ and H₂S in mg/L, Rn in nCi/L and all others in meq/L)

Parameter	Afra springs	Weidaa spring	Ibn Hammad Upper	Himma spring Yarmouk River	North Shuna Therm. Spring	Manshiya well 2
EC	545	610	847	1320	863	735
pH	6.96	6.5	6.8	6.98	7.06	6.34
Ca^{+2}	2.17	3.80	4.2	6.21	4.08	3.4
Mg^+	1.40	1.21	2.42	2.7	3.66	3.25
Na^+	1.48	1.68	2.57	5.87	3.31	0.91
K^+	0.02	0.21	0.10	0.44	0.21	0.06
Cl^-	1.68	2.48	4.51	6.03	2.81	1.03
SO_4^{-2}	1.35	1.77	2.44	3.45	1.76	0.90
HCO_3^-	1.98	3.00	3.05	5.54	6.38	7.33
NO_3^-	0.00	0.0	0.0	3.10	1.24	0.0
T °C	45.9	33	37	41.4	52.7	54.3
Rn	7.38	2.86	5.9	31.46	2	16.5
H_2S	0.0	0.0	0.0	2.8	5.83	0.2

In Summary the Natural Water Resources Quality Can Be Described As Follows

1. All floodwater resources in Jordan have qualities which allow them to be used for all common purposes. For domestic and certain industrial uses, they may require filtration to remove silt particles and chlorination as a precautionary measure.

2. All renewable water resources (in their natural states) along the highlands and on the plateau are also suitable from a qualitative point of view for all common uses.

3. The natural groundwater qualities in the intermediate and deep aquifers with a free water table are suitable for all common uses, without any treatment except chlorination as a precautionary measure.

4. The natural groundwater qualities in the Jordan Valley are also suitable for all common uses except when they are in contact with the salty deposits in the valley, especially the Lisan Formation or contaminated by irrigation return flows.

5. The confined deep aquifer and the confined B2/A7 and A4 aquifer portions contain water with high salt concentrations, CO_2, H_2S and radon gases and have temperatures exceeding the ambient temperature by more than 5 °C reaching a maximum of 64 °C. Hence, this water is generally not suitable for common uses, except after treatment which may incorporate desalination.

6. The slowly rechargeable aquifers of north-east Jordan, Sirhan, Hamad and Azraq basins contain generally high salt concentrations, of 1500–5000 µS/cm, which classifies them as brackish waters. They are only conditionally suitable for common uses after certain treatments.

Table 4.7 Concentration ranges of radioactive elements in minerals and rock

Mineral	Concentration range in ppm
Quartz	0,1–3.5
Mica	0.1–40
Feldspars	0.2–8.2
Accessory minerals	
Apatite	10–90
Monazite	200–850
Zirconium	100–4500
Xenotime in siliciclastic sedimentary rocks	300–12,500
Igneous rocks	**Average ppm**
Acidic rocks	4
Intermediate rocks	2
Basic rocks	0.9
Ultra-basic rocks	0.03
Metamorphic rocks	**Concentration range in ppm**
Gneiss	3–7
Amphiboples	2–3
Schist	1–3
Marble	0.1–0.3
Quarzite conglomerate	0.3–4.0

4.2   Natural Radioactivity in the Water Resources

Naturally most rock formations are radioactive but in different degrees. Igneous rocks which are the source of all other rock types contain radioactive minerals such as uranium, thorium, potassium, strontium and others. Their daughter elements are also mostly radioactive with half-life times ranging from milliseconds to millions of years (Table 4.7). In Jordan, the original source of radioactive minerals is the granitic basement complex cropping out in Aqaba Mountains and extending under Jordan's territories with increasing depths in a north-easterly direction, e.g. at around 6 km under the Azraq area (Fig. 5.1). Throughout the geologic history of Jordan this granitic basement, which used to extend further south and south-west into Saudi Arabia and Egypt before the opening of the Red Sea, delivered the weathering products of sand, silt, clays and gravels including radioactive minerals which were deposited on Jordanian territory overlying the extensions of the basement complex further north, north-east and north-west of Aqaba. These weathering products were deposited as terrestrial clastic sediments together with the radioactive minerals, which had no opportunity of being removed to the open sea or further away elsewhere. Hence, the mostly clastic sediments from the Cambrian through the Lower Cretaceous contain radioactive minerals. The thickness of these clastic sediments ranges from zero in Aqaba granitic basement to more than 3500 m in north Jordan.

Table 4.8 Concentration ranges of radioactive elements in the sedimentary rocks, worldwide and in Jordan

Sedimentary rocks	Concentration range in ppm	Concentration range in Jordan in ppm
Limestone	0.1–1.3	Up to 2.8
Quartz sand	0.6–2.2	Sandstone aquifer complex 180–526
Clays	3.0–4.0	2.5–8.6
Bituminous rocks	1.4–1200	115–850
Phosphates	10–1500	40–250
Thermal springs in sedimentary rocks		60–170

In addition, during the different ingressions of the Tethys over Jordan or parts of it, marine sediments such as limestone, shale, and phosphate rocks, among others, were deposited together with their radioactive minerals originating from their source rocks of the granitic basement and its overlying sediments. The thickness of these sediments reaches around 2500 m in north Jordan. Depending on the sedimentation environment, the sedimentary rocks contain varying concentrations of radioactive minerals, which are generally concentrated in rocks rich in organic matter, such as oil shale and phosphate rocks (Table 4.8).

Triassic and Jurassic rocks which intercalate into the geologic column of Jordan, starting at the latitude of the middle of the Dead Sea; the Zarqa Ma'in area and north of it are also composed of terrestrial and shore deposits of siltstone and sandstone and further north, marine clay and limestone are common. The clastic terrestrial and shore sediments also contain radioactive minerals originating from the granitic basement complex or from the reworking of the rock deposits, which pre-date the Triassic. The marine Triassic and Jurassic sediments further north were deposited in a reducing environment favoring the accumulation of heavy and trace minerals including those of radioactive uranium and thorium, among others.

The Upper Cretaceous marine rocks of the Nodular, Echinoidal and Massive Limestone Formations composed of dolomite, limestone clay, marl and shale contain in their original composition low concentrations of radioactive minerals. The uppermost Upper Cretaceous and Tertiary rocks composed of phosphate and oil shale contain high concentrations of radioactive minerals mostly accumulated in the organic matter of these rocks.

The leaching of phosphate and oil shale rocks results in water loaded with radioactive minerals. Once these come into contact with the atmospheric air they precipitate in and on other rocks' porosity spaces, joints and fractures. This is the case with the uppermost Upper Cretaceous limestone beds and recent surface terrestrial sediments in central Jordan (about 20 km south of Amman International Airport), where the uranium deposits are found overlying the oil shale deposits.

The thickness of Upper Cretaceous and Tertiary rocks which contain radioactive minerals in their composition reaches in north and east Jordan around 550 m, added to that are the deep sandstone aquifer complex of about 3500 m. The total rock thickness of the geological column containing the radioactive minerals sums up to about 4000 m. This constitutes one of the best' exploited aquifers in Jordan.

The Upper Cretaceous rocks containing low concentrations on radioactive minerals are generally considered as aquicludes or badly developed aquifers with very limited yields.

Considering the information given above, almost all relevant exploited aquifers in Jordan contain radioactive minerals which are released into the groundwater in them. The concentrations in the groundwater then reflect the groundwater conditions, confined or unconfined, the aquifer rock matrix content of radioactive minerals and the type of mineral in which the radioactive metals are incorporated.

Therefore, Jordan's viable groundwater resources contain naturally higher concentrations of radioactivity. Nonetheless, exposures in the worst cases do not exceed 0.7 milli Sievert/year (mS/year), and generally it is below 0.5 mS/year.

The WHO report 2013, which recommends exposure radioactivity through water of less than 0.1 mS/year emphasizes the fact that: "No international standards for drinking water quality are promoted for adoption."

The report also states that: "It is essential that each country review its needs and capacities in developing its regulatory framework."

The main reason being: "The advantage provided by the use of a risk-benefit approach, qualitative or quantitative, in the establishment of national standards and regulations,"

According to the report, the guidelines provide a "scientific point of departure" for national authorities to develop drinking water regulations and standards appropriate to the national situation. (In Australia the guidelines promote a concentration of up to 1 mS/year)

In particular, Chapter 9 of the WHO report, which discusses radiological aspects, emphasizes that: "Screening levels and guidance levels are conservative and should not be interpreted as mandatory limits." And exceeding a guidance level should be taken as a trigger for further investigation, but not necessarily as an indication that the "drinking-water is unsafe."

The WHO report also emphasizes the fact that background radiation exposures vary widely across the Earth, but the average is about 2.4 mS/year, with the highest local levels being up to 10 times higher without any apparent health consequences.

Therefore, the amount of radioactivity contributed by drinking water represents a small addition to background levels (Fig. 4.1).

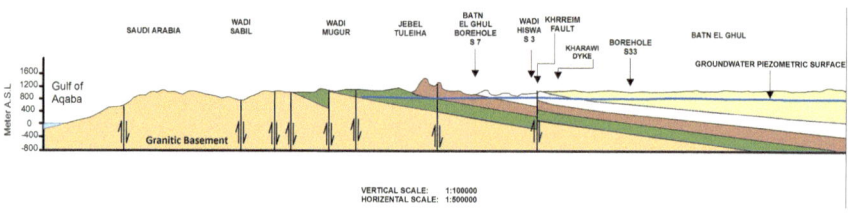

Fig. 4.1 Simplified geologic north-south cross-section with the granitic basement underlying all geologic units in Jordan and being the source of radioactivity of all other rock units

4.3 Water Quality As Affected by Human Activities

Generally, water resources are exposed to pollution factors which affect their qualities; these include human activities like the disposal of solid and liquid wastes of urban and industrial areas, the use of biocides and fertilizers in agriculture, and the return flows to surface and groundwater resources from irrigation water, as well as the over-exploitation of groundwater resources.

4.3.1 Pollution Sources

For the case of pollution in Jordan, Table 4.9 shows the producers, the sites, the affected environmental element and the affected group.

The drastic increase in Jordan's population, improving living standards and expanding industrialization have demanded increases in water supplies, which have in turn produced increasing amounts of waste water and irrigation return flows, and led to the total use of all available surface water resources and to the over-exploitation of aquifers. The results are deteriorating surface and groundwater qualities.

The following describes the major polluted areas and discusses the causes of water quality deterioration.

A. Domestic Wastes

The wastes produced in households consist of waste water, solid wastes and gases emitted into the atmosphere. In Jordan, the two main waste types which result in water pollution are waste water and solid wastes.

Domestic Waste Water
The waste water of households is disposed of in two ways:

- Cesspools and septic tanks.
- Collection via sewerage systems and treatment in waste water treatment plants.

Each of these has its impacts on the local surface and groundwater resources.

Human sewerage contains organisms that cause a variety of diseases such as cholera, typhoid, and dysentery. It also has relatively higher concentrations of nitrates and phosphates and higher salinity than the supply waters. If not dealt with properly it leads to disease and death, contaminates surface and groundwater resources, produces odor and may pollute soils, threatening the health of cattle and contaminating crops. Therefore, it represents a significant problem for all communities and countries, because abating its possible impacts requires relatively large investments, treatment costs, know-how and safe reuse schemes.

Generally, waste water contains higher dissolved salt concentrations than those used to supply water. The increase in salt contents results from the increased use of chemicals in households and their evaporation and is a function of the per capita water consumption. The lower the consumption, the higher the concentration of

Table 4.9 Water quality problems in Jordan

Type of pollution vs. its effects	Deterioration of water quality as result of over pumping	Deterioration waste water quality due to low water consumption	Inadequate waste water treatment, biocides and pharma residues	Waste water low coverage	Irrigation return flows	Solid waste disposal	Industrial waste water
Affected area	Overall north, partly central and southern Jordan	Whole country	Whole country	Overall where cesspools are in use	All over where irrigation is practiced	All over where solid wastes disposal is practiced	Amman, Zarqa, Balqa, Aqaba, Irbid etc.
Affected environmental element	Ground water and surface water	Water, soils, plants and food	Water, soils, and food	Ground and surface water	Ground and surface water	Soils, surface and groundwater, air quality	Surface and ground water, plants, soils
Causes	Inadequate water quality, deteriorating surface and ground water	Polluted irrigation water, soils, ground and surface water	Deteriorating soils, surface and ground water qualities and human and animal health	Human and animal health, soils, surface and ground water qualities and quantities	Deteriorating soils, surface and ground water qualities and lower land productivity	Increasing salinity of soils, ground and surface water and trace elements and pharma residues	Contamination of surface and ground water with trace elements, salinity industrial chemicals

salts. Waste water treatment can remove a variety of pollutants and substances but salts can only be removed by very expensive techniques, such as desalination. Hence, they are not removed from the waste water during treatment.

Crops grown in Jordan; vegetables and fruit trees produce the largest yields per unit of land if the salinity of the water is less than 1200 µS/cm. If salinity increases beyond that, the result is a reduction in crop production. Therefore, the use of treated waste water with salinity higher than 1200 µS/cm may result in crop reductions and declines in productivity.

Cesspools

Many towns, villages, settlements, higher education centers, military camps … etc. are still not connected to sewerage systems and waste water treatment plants. Even in sewered towns there are households that still use cesspools. In addition sewer pipes are not totally secure, therefore, untreated waste water leaks from them.

Generally, cesspools are not designed to hold the waste water to be transported later on to a waste water treatment plant. The bottoms and walls of cesspools are normally permeable, especially in those types of rocks generally found in the highlands, such as karstic limestone, fractured chert and silicified limestone, sandstones or disintegrated shale.

These conditions allow the infiltrated sewerage water to reach the underlying water bodies, especially those phreatic ones and those found at shallow depths.

In the following, examples of local pollution caused by infiltrating waste waters are given.

– Amman Area

Although Amman possesses a sewerage system and a sewerage treatment plant, some of its outskirts still lack the connection to a waste water treatment plant. In addition, sewerage pipes are not very secure, therefore they leak into the surrounding soils and rocks. The city is built on limestone, chert and silicified limestone; all of these are highly permeable and build aquifers in the area. The waste water collected in cesspools and that which leaks from the sewers infiltrates into the rocks and reaches the groundwater bodies resulting in their pollution. Examples of this are Rasel Ain and Rusaifa springs. The pollution by waste water is reflected in the nitrate contents of the springs of 72 and 60 mg/L compared to a natural concentration in phreatic aquifers in Jordan of 15–20 mg/L. Already in 1966 Rasel Ain spring had a nitrate concentration of 30 mg/L, almost double the natural concentration. The rapid increases in the concentrations of sodium and chlorides indicate also the effects of waste water. Worth mentioning here is that, in the up-gradient areas of these springs, industrial and irrigation activities are absent, which excludes them as a cause of the pollution.

– Zarqa-Marka Area

This area also possesses a sewerage system and numerous industrial waste water treatment plants. The rocks underlying the area consist of chert, phosphate, limestone; all of these are permeable and form the areas' aquifers. Leaky cesspools and sewers have also resulted in increases in salinity, sodium, chloride and especially nitrate concentrations.

The EC natural values ranged from 500–700 μS/cm; now they are two- to two and a half- fold that value. The strong increase is in the nitrate content; 3–4 times the natural concentration. Zarqa municipality well can only be safely used for municipal supply after mixing with a better quality water to lessen its salt content and nitrate concentration in order to comply with the Jordanian drinking water guidelines.

– *Baqa'a Area*

Baqa'a Refugee Camp and the surrounding villages possess a sewerage system and a waste water treatment plant. But many houses constructed on the surrounding mountains possess cesspools and no central sewerage system. The aquifer underlying the area is the Kurnub Sandstone, which is phreatic with a water table of up to 100 m below ground level. The surrounding mountains are built of highly fractured disintegrated limestone and shale which allow infiltration into the underlying sandstone aquifer. Therefore, infiltrating sewerage water reaches the groundwater body and pollutes it. An example on this is well AL 1395 originally (1975) with a nitrate concentration of 13 mg/L and EC value of 660 μS/cm increasing to 58 mg/L and around 1220 μS/cm in 2005 (Figs. 4.2 and 4.3). The fivefold increase in the nitrate concentration compared to the twofold increase in salinity indicates a sewerage type of pollution. The further surroundings of the well area, especially its up-gradient parts are devoid of any major industry or agricultural activity, hence excluding these as sources of pollution.

– *The highland towns of Hartha, Irbid, Salt, Na'ur and Karak*

Karak, Salt and parts of Irbid possess sewerage systems and waste water treatment plants. Nevertheless, cesspools and leaky sewers still infiltrate through the karstic aquifers working up the highlands to the groundwater bodies. Limestone,

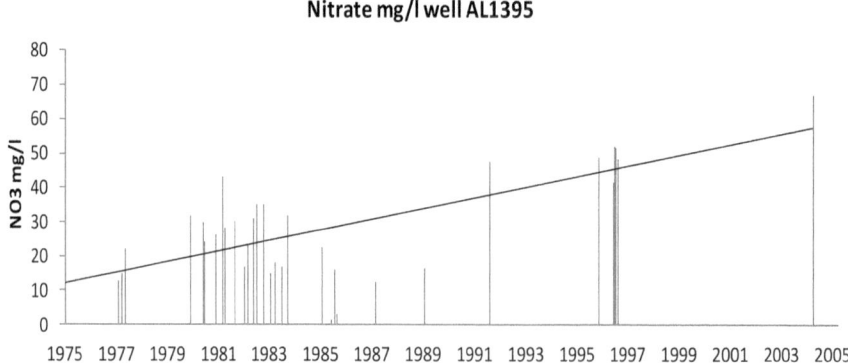

Fig. 4.2 Increases in NO₃ concentrations in a well in Baqa'a area as a result of sewerage water infiltration in the area

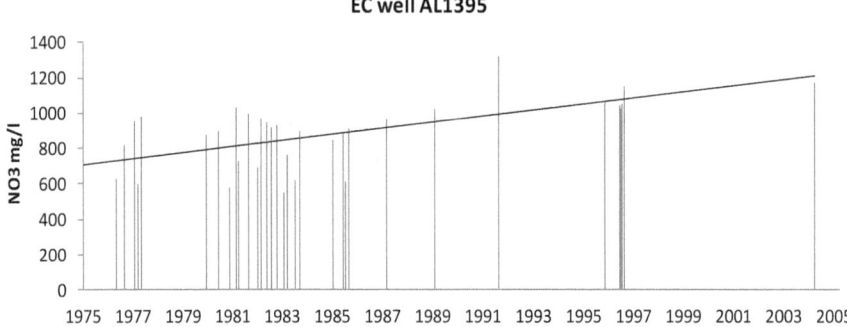

Fig. 4.3 Increases in EC values in a well in Baqa'a area as a result of sewerage water infiltration in the area

Table 4.10 Recent analyses of spring water of towns along the highlands possessing centralized sewerage systems but still have cesspools (mg/L)

Parameter	Hartha	Irbid	Salt	Naur	Karak
EC µs/cm	467	505	1229	761	915
pH	8.26	7.00	7.47	7.35	7.52
HCO_3^-	232.4	232.41	301.34	283.04	308.05
CO_3^{-2}	0.00	0.00	0.00	0.00	0.00
Hardness	222.0	210.00	455.00	303.00	343.00
Ca^2	68.7	62.12	101.20	79.56	92.38
Mg^2	12.3	13.38	49.13	25.41	27.36
Na^+	13.8	17.48	60.95	20.93	43.47
K^+	1.56	0.78	12.90	2.74	22.29
Cl^-	19.53	22.37	134.90	49.70	74.55
SO_4^{-2}	6.72	11.52	120.48	15.84	36.96
NO_3^-	28.01	22.46	61.79	31.87	79.15

silicified limestone, chert and shale are the surface rocks underlying these towns. In Salt, Karak and Hartha the limestone of the Upper Cretaceous and Lower Tertiary are highly karstified. Cesspool contents of waste water infiltrate rapidly into the groundwater bodies. Some of the cesspools especially in the Hartha area are not able to hold any water.

The original water quality in all these areas showed EC values of 500–700 µS/cm and nitrate contents of 15–20 mg/L. Table 4.10 shows the moderate increases in the EC values compared to 2–7-fold increases in the nitrate contents. These areas are devoid of any major industry or intensive agricultural activity. Hence, pollution is related to household types of waste water (Table 4.11).

Table 4.11 Chemical analyses of spring water in Jordan

Water source	Ras el Ain Sp. Amman		Rusaifa Sp.	Pepsi W. Rusaifa	Zarqa Municp. W.	Shami W. Baqa'a		Na'ur Sp.	Sara Sp. Karak	Jadour Sp. Salt	Kufr Assad W. Irbid	Nuayma Municip. W.		Um Irshed Sp. Hartha
	1966	2012	2012	2013	2013	1981	2015	2014	2015	2013	2014	1975	2015	2015
EC μS/cm	520	950	915	1120	1780	600	740	1160	706	830	1350	590	630	770
Ca^{+2} meq/L	3.71	5.95	2.5	5.4	5.2	3.4	4.0	6.26	3.8	4.51	8.80	3.2	3.5	4.65
$Mg + 2$ meq/L	1.15	1.45	1.7	1.75	1.25	1.9	2.13	1.94	1.40	2.21	2.55	1.9	2.0	2.41
Na^+ meq/L	0.3	2.09	2.47	3.70	8.70	1.13	1.12	3.1	1.75	1.81	2.95	0.91	0.91	2.1
K^+ meq/L	0.06	0.12	0.10	0.10	0.12	0.1	0.19	0.26	0.5	0.61	0.15	0.04	0.06	2.23
Cl^- meq/L	0.67	2.54	3.84	4.	9.83	1.16	1.30	3.49	0.5	1.65	3.22	0.89	1.19	3.29
SO_4^{-2} meq/L	0.75	0.52	0.42	0.37	3.02	0.71	0.65	0.46	0.52	1.21	3.80	0.14	0.18	1.07
HCO_3^- meq/L	4.22	5.4	1.82	4.54	5.02	4.42	4.20	5.5	3.70	4.62	6.20	4.57	4.57	3.77
NO_3^- mg/L	30	72	60	58	67	13	78	87	55	94	70	20	48	198

Reference

Salameh E (1996) Water quality degradation in Jordan. Friedrich Ebert Stiftung, Amman and Royal Society for the Conservation of Nature, Amman, 179 p

Chapter 5
Waste Water Treatment

5.1 Introduction to Waste Water Treatment (WWT)

Water use in household, industry or agriculture produces waste water which cannot be used again for the same purpose which caused its deterioration. Municipal waste water is mainly composed of wastes from cooking/kitchen activities, body and clothes washing. Industrial waste water contains a very large variety of chemicals some of which are hazardous to humans, animals, plants and nature. Domestic waste water is also hazardous because it contains disease-causing organisms.

The proper treatment and safe disposal of waste water are of great importance for two main reasons:

1. Removal of disease-causing organisms thus stopping the spread of communicable diseases.
2. Disintegration of the organic matter and prevention of pollution of surface and groundwater resources.

In countries rich in water resources and rivers, the waste water is treated before disposal into watercourses resulting in its dilution. In water-scarce countries with high population density treated waste water represents an additional water source which can be used for some purposes according to its quality. In such countries the agricultural sector is the main candidate for such reuse purposes.

The water scarcity in Jordan forces the reuse of treated waste water in agriculture. To protect the environment and to enable the reuse of the treated waste water in irrigation many waste water treatment plants were constructed in Jordan to treat the municipal waste water of all cities and towns. At present, some 120 million m^3/year of treated waste water are reused to irrigate at least 240,000 dunums with alfalfa and various trees such as olive and apple.

If the waste water is treated properly, the effluents can be reused to irrigate different crops, in the knowledge that municipal waste water is rich in nutrients and can from this point of view result in higher agricultural productivity (Fig. 5.1).

Sections 5.1, 5.2.1 and % 2.2 are based on Salameh (1996) with modified and updated figures and data from MWI (2016).

© Springer International Publishing AG, part of Springer Nature 2018
E. Salameh et al., *Water Resources of Jordan*, World Water Resources 1,
https://doi.org/10.1007/978-3-319-77748-1_5

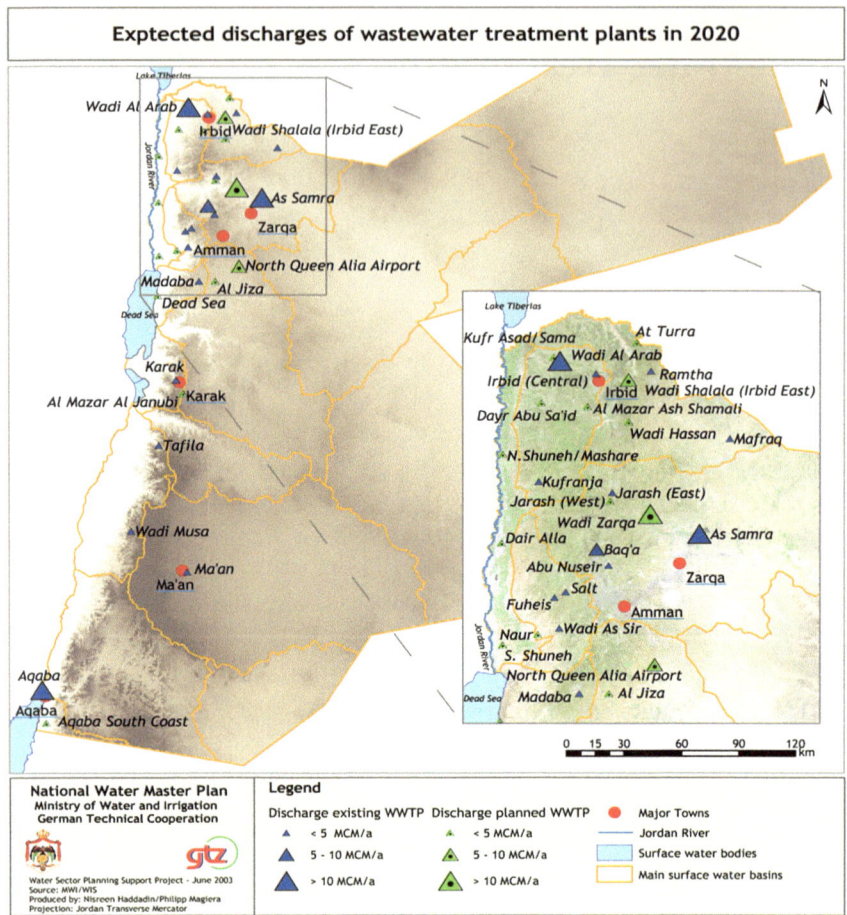

Fig. 5.1 Expected discharges of waste water treatment plants in 2020

5.2   Methods of Waste Water Treatment

Municipal waste water can be treated in different ways using a variety of methods. The most appropriate and fastest treatment methods for large communities are mechanical treatment plants. More natural methods such as waste stabilization ponds, oxidation ditches and aerated lagoons can be used for small communities and for small amounts of waste water.

5.2.1   Conventional Waste Water Treatment

Conventional waste water treatment comprises the following stages:

1. **Preliminary treatment**; in this stage, large suspended solid particles are removed through screening.
2. **Primary treatment**; this stage uses gravitational separation for suspended materials to settle at the bottom of treatment pools. This stage aims at clarifying the waste water from its suspended solids, which reduces the load of the second stage of treatment namely; the secondary treatment stage.
3. **Secondary or biological treatment stage (bio-filtration or activated sludge)**; the liquid resulting from the primary sedimentation is treated in a bio-filter or in activated sludge pools.

In the case of bio-filtration the settled sewerage is spread over a 1.8 m deep bed of coarse aggregates. The liquid trickles down to the surface of the aggregates, developing a microbial film where bacteria oxidize the sewerage as it flows over.

In the case of activated sludge, the settled sewerage is moved to an aeration tank, where the mechanically supplied oxygen oxidizes the sewerage.

Sludge resulting from the primary and secondary treatment stages is anaerobically digested and can be reused as fertilizer.

Conventional treatments can achieve high BOD removal in very short retention times, which makes them very efficient. They are mainly used where land is not adequately available and where evaporation losses are high due to the high temperatures of the surroundings.

Conventional waste water treatment plants are expensive to construct and maintain and require skilled operators. They reach high efficiency in BOD removal, but pathogen removal is very low making chlorination of treatment effluents imperative from a health point of view. But chlorination of treated waste produces trihalomethanes which are carcinogenic agents.

5.2.2   Less Conventional Methods of Waste Water Treatment

A. *Oxidation Ditches*

Oxidation ditch treatment represents a modification of the conventional activated sludge process. The screened raw sewerage is diverted into the oxidation ditches for long retention times of commonly 0.5–1.5 days; in the case of solid particles it is 20–30 days. Mechanical rotation is needed to aerate the sewerage with a BOD_5 removal rate of more than 95%.

B. *Aerated Lagoons*

Aerated lagoons are conventional units of activated sludge, but operated without sludge return. They are waste stabilization ponds with added mechanical aeration to supply algae with additional oxygen. They achieve BOD_5 removal of more than 90% but with long retention times of 2–6 days.

C. *Waste Stabilization Ponds*

Waste water is treated by waste stabilization in large, shallow basins (1–2 m in depth) separated from each other by embankments. The raw waste water enters the ponds and flows into successive ponds where in each pond a stage of natural treatment by the action of bacteria and algae takes place. Three basic treatment types are incorporated: anaerobic, facultative and maturation.

In this type of treatment mechanical aerators to produce oxygen are not needed and oxygen is symbiotically produced by algae. Sludge recycling is not required. Treatment in such ponds requires a long retention time, but under the appropriate conditions they produce good effluent quality with a BOD_5 removal >95% strong reduction in faecal coliforms.

In 1945 the first waste water treatment plant in Jordan was built to serve Salt City which suffered from pollution of its springs' water as a result of seepages from cesspools and cesspits into these springs which emerge in the area below Salt City along Wadi Salt. The inhabitants of the city at that time began to suffer from waterborne diseases and it was concluded that the pollutants originating from waste water return flows into the springs' water were negatively affecting the water quality in the area as well as transmitting various diseases.

The treatment in this waste water treatment plant was of the extended aeration type. Upon its construction together with the sewerage system in the city, cesspools were abandoned and springs and groundwater qualities improved. The treatment plant functioned efficiently so that effluent was used in unrestricted irrigation along Wadi Shueib without any signs of environmental or health hazards.

The pollution of drinking water resources also occurred in Amman, Irbid and other towns and cities in Jordan, which led to the construction of waste water treatment plants from the early 1960s onward. The first treatment plants were of the trickling filter and activated sludge types and have functioned efficiently without any major problems except when they were hydraulically or biologically overloaded (Table 5.1).

Contrary to that, treatment in waste water stabilization ponds was introduced in Jordan in 1984; Khirbet-es-Samra treating the waste waters of Amman was applied also to serve the towns of Aqaba, Ramtha Tafilah, Maan and Mafraq. All these treatment plants failed their purpose of producing treated effluent qualities which could cope with the environment and further usage as irrigation, and they were replaced after 2007 by mechanical waste water treatments, that produced better effluent qualities. Examples of the quality of stabilization plants' effluents is given in Table 5.2.

The quality of waste water treatment plants' effluents at present are given in Table 5.1. It shows that the effluents are within the Jordanian Standards for waste

Table 5.1 Average composition of treated waste water of some Jordanian waste water treatment plants for the year 2015 (mg/L)

Plant	PO$_4^{-3}$	NO$_3^-$	T-N	NH$_4^+$	TDS	TSS	COD	BOD$_5$	[a]pH
Irbid	6.9	1.4	93.7	80.2	1075	144	326	64	7.54
Fuheis	12.5	6.1	19.8	16.2	896	82	130	39	7.36
Wadi Arab	8.9	0.8	68.1	65.5	991	33	77	23	7.67
Abu Nuseir	5.9	18.6	8.6	0.7	826	11	37	11	7.11
Samra	4.8	47.1	17.5	2.9	928	18	58	9	7.59
Baqa'a	7.8	12.4	39.4	27.3	969	33	114	37	7.63
Wadi Sir	13.6	1.0	85.4	61.8	828	32	142	61	7.52
Aqaba	1.7	4.1	4.9	3.1	613	6	23	5	7.24
Wadi Hassan	6.3	18	11.0	3.0	1136	14	59	20	8.07
Ma'an	0.8	15.6	8.3	1.7	909	15	47	6	7.51
Ramtha	15.3	43.1	83.2	69.0	1368	28	90	4.	7.54
Madaba	1.4	5.1	35.2	34.3	1124	19	67	25	7.55

[a]Without unit

Table 5.2 Main pollution parameters in the effluents of some waste water treatment plants working on stabilization pond principles

Parameter year's average	Khirbet as Samra	Mafraq	Aqaba	Ma'an	Ramtha
BOD$_5$ mg/L	144	272	74	170	290
COD mg/L	356	432	323	353	595
TSS mg/L	126	181	3	211	160

water treatment effluents to be discharged along wadis and to be used, mixed with flood and base flows for the irrigation of crops and in some cases for restricted crops.

Figure 5.2 gives the differences between effluents of stabilization pond treatments and the mechanical treatments to indicate the radical improvements in waste water treatment after substituting stabilization ponds by mechanical treatment.

5.3   Summary of Domestic Waste Water Treatment Plants

Different guidelines are given in the literature concerning the effluent of treatment plants; they generally depend on the environmental conditions in the downstream areas, the next use of the waters and the environmental impacts of their discharges especially on the recipient water bodies. In Middle Europe, 30 mg/L BOD$_5$ are recommended as an upper limit. Elsewhere this concentration can go up to 100 mg/L if environmental conditions allow it.

For Jordan, with its extreme water scarcity and high value of water, a BOD$_5$ concentration of 30 mg/L should be targeted, especially because experience with these treatment plants' effluents indicates severe environmental impacts on surface and groundwater.

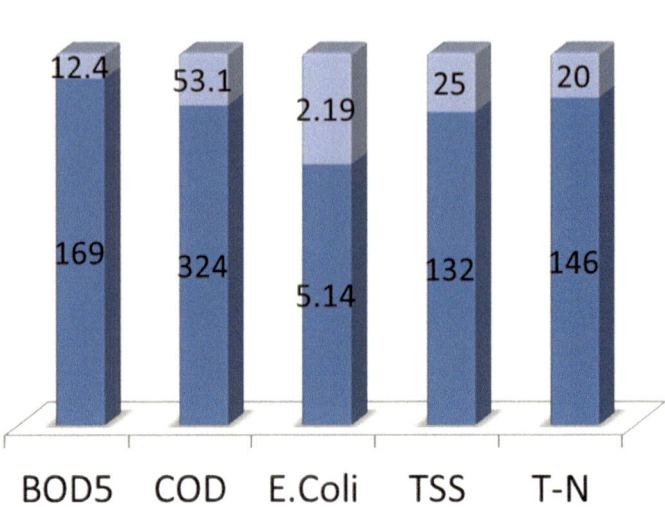

Fig. 5.2 Comparison between the old treatment plant (stabilization ponds) and the new treatment plant (mechanical treatment)

Most of the treatment plants working on mechanical principles produce an effluent with BOD_5 concentrations of less than 30 mg/L. The treatment in these plants involves mechanical components, trickling/aeration, activated sludge … etc. Only the old treatment plants of Irbid, Baqa'a, Fuheis and Wadi Sir produce an effluent with higher BOD_5, due to hydraulic overload and to higher organic matter concentration in the incoming waste water.

After introducing mechanical waste water treatment instead of stabilization ponds the effluents of treatment plants improved radically and the treated waste water became of good quality safe for reuse in the irrigation of different crops.

Solid Waste Disposal

The impacts of solid waste disposal on surface and groundwater resources are well recognized and documented in the literature (Farquhar 1989). The problem is that once these impacts are recognised in the composition of a groundwater body manifested in the detection of pollutants, the repair is very difficult or even impossible.

Generally, municipal solid wastes contain only limited numbers and amounts of dangerous chemicals. But, in certain cases where separation of wastes is not practiced, even dangerous chemicals, such as insecticides, pesticides, all types of medicaments, batteries, paints, and mineral oils among others, reach the solid waste disposal sites, become dissolved and leak to ground and surface water bodies.

Of concern here in Jordan are the solid waste disposal sites distributed all over the country. The sites of Amman, Zarqa, Salt, Madaba, Ukheider, Karak, Hartha, and Aqaba among others were not exposed to sophisticated environmental impact assessment in advance. Hence, they do not fulfil present-day environmental criteria.

In addition, most of the sites are located along wadis or depressions of abandoned quarries or karst holes, forming therefore a direct threat to the ground and surface water resources.

Solid waste in Jordan consists of about 49.5% food waste, 25.9% cartons, 6.5% plastics, 3.3% glass, 2.5% metals and the rest of natural materials; wood, earth, stones, construction refuse etc. (El-Natour 1993).

At Amman solid waste disposal site, the waste is brought by trucks. Unseparated, it is deposited, consolidated and covered by a few centimeters of soil. Such a procedure only reduces the entry of air, and prohibits immediate burning and combustion; it also does not spoil the landscape, compared to open uncovered dumping sites.

Wells in the upstream area of the solid waste disposal site (AL 2106) show relatively constant concentrations of the different parameters. Wells lying down-gradient of the site show, contrary to that, increases in the concentrations of the different parameters (AL 1326). Further down-gradient wells: AL 1303, AL 1318 and AL 1302 show less expressed increases in the concentration of the different parameters as a result of the divergence of the pollution plumes.

The composition of the leachates originating from the solid waste disposal site of Amman is given in Table 5.3. The table shows also the composition of the liquid waste pool in Wadi El-Kattar in the immediate neighborhood of the solid waste disposal site.

In practice, many years may elapse before the solid waste disposal site reaches the field capacity to generate leachate, which are almost entirely, 98%, composed of calcium, magnesium, sodium chloride, sulphate, nitrogen compounds (such as ammonia, nitrate and nitrite) and bicarbonates. Also time is required before peak concentrations of the various parameters are reached. However, once this peak is reached, it continues at the same level for tens or hundreds of years after the end of the disposal.

Poorly biodegradable, low-solubility contaminants appear and reach peaks way after the normal chemical constituents mentioned above. Hence, they persist longer and affect the water resources for long periods of times, tens to hundreds of years (Farquhar, 1989).

In the other areas of Jordan solid waste disposal sites are not better off than the one in Amman. Serious impacts on the surface and groundwater resources are unavoidable if not already present.

Industrial Effluents
As regards treatment, waste water effluents of industries in Jordan can be categorized into the following:

A. Treated effluents.
B. Untreated effluents.
C. Semi-treated effluents.

By law, industries in Jordan are obliged to treat their waste water before releasing it into the environment. This has forced all major industries to treat their waste water before reuse, or before discharge to domestic waste water treatment plants or into the environment. At present 90% of all industries treat their waste water in their own

Table 5.3 Analyses of the solid wastes leachates, and the liquid waste pool, south-east of Amman

Parameter	Temp °C	EC mmoh/ cm	pH	NO_3^- meq/L	HCO_3^- meq/L	CO_3 meq/L	SO_4^{-2} meq/L	Cl^- meq/L	Ca^{+2} meq/L	Mg^{+2} meq/L	Na^+ meq/L	K^+ meq/L
Old dump	14.6	138.0	8.46	12.09	6.09	0	11.5	150	7.3	8.6	142	23.22
New dump	14.4	5.500	8.10	0.062	4.48	0	1.9	34.1	2.4	3.2	37.6	1.18
Waste pool	15.6	18.560	8.68	89	3.50	0	4.88	158.9	7.71	9.77	154.3	2.65

plants. Most of the rest of the industries are allowed to discharge their waste water to municipal waste water treatment plants because it generally resembles household waste water. In general effluents of industrial treatment plants comply with regulations and standards, but some do not, and very few produce semi-treated waste water (Salameh 1996).

As regards the method of effluent disposal, industrial waste water can be categorized into:

1. Added to municipal waste water treatment plant
2. Directly discharged into the environment, especially along wadis
3. Reused for irrigation within the premises of the industry
4. Collection in special pool allowing its evaporation.

Around 44% of the industries discharge their pre-treated effluents to domestic waste water treatment plants. Others are not allowed to do this because their effluents contain substances which pose a great threat to the treatment processes of domestic WWTP. These industries discharge their effluents to the nearby wadis. Some industries, such as the yeast industry and some oil refineries and others use the effluents in local irrigation.

Some industries with effluents of a few cubic meters per day transport their waste water by trucks to pools and spreading basins where they infiltrate or evaporate.

According to the quality degradation parameters contained in industrial effluents, industrial waste water can be classified into the following:

(a) High salinity effluents.
(b) High organic loads effluents.
(c) High trace elements effluents.
(d) High contents of other specific substances such as phosphates and ABS.
(e) A combination of any group of (a) to (d).

The majority of industries have effluents with high salinity or high organic load or both. A few of them have high trace element concentrations, and very few have high concentrations of specific parameters.

Trace Elements
High concentrations of trace elements are found in the effluents of the industries listed in Table 5.4. These effluents end up in the surface and groundwater bodies and in the soils. Locally, they may threaten surface and groundwater bodies and may if connected to domestic waste water treatment plants, kill the active bacteria and algae and hence stop the biological activities involved in waste water treatment (Salameh 1996).

The destination of these trace elements in the Amman/Zarqa area is King Talal Dam (KTD). Therefore, the inflows to KTD, whether base or flood flows, contain some trace elements dissolved in the water itself and attached to or bound into the suspended particles which are transported by the water. In KTD the particles settle to the bottom of the reservoir, but they are partially re- released into the water due

Table 5.4 Examples of industries with high trace element concentrations in their treated effluents (mg/L)

Industry	Av. discharge m³/ day	Trace elements								
		Fe	Mn	Zn	Cu	Cd	Cr	Ni	Pb	Hg
Tanning	370	0.43	0.16	0.32	0.120	0.052	1.92	0.25	0.25	0.001
Yeast	330	4.20	2.00	0.18	0.072	0.020	0.08	0.23	0.072	0.0008
Battery	12	16.30	6.50	0.45	0.620	0.020	0.08	0.25	6.32	0.005
Chemical ind. (detergents)	8	0.50	0.32	0.21	0.09	0.090	0.21	0.26	0.32	0.15
Sulfo-chemicals	62	3.50	1.40	0.16	0.090	0.700	0.23	0.25	0.35	0.0009

to the eutrophication processes taking place there. Afterwards they are discharged along Wadi Zarqa and the water is used for irrigation in the Jordan Valley area.

Until now these trace elements have not represented a threat to downstream irrigational uses. But, a combination of factors such as less irrigation water per unit of land, water depletion of oxygen and high water salinity may activate these trace elements and make them available to plants which would weaken the these plants' resistance to diseases and may result in crop failure.

The total loads of trace elements to the whole water system in the most industrialized area in Jordan, the Amman – Zarqa area, is relatively small and it does not represent any threat to the system as a whole. But on a local scale some trace elements such as lead (battery industry) or mercury (chemical industries or chlorine filling plants) represent an acute danger to surface and groundwater resources. Therefore, these industries should treat these trace elements adequately before any waste water releases into the surroundings. The outlet water of KTD contains some high concentrations of trace elements such as lead, 15–40 µg/L, nickel, 50–100 mg/L, chromium, 50–62 mg/L, cadmium, 0–20 mg/L, zinc, 70–88 mg/L, Cu, 2–12 mg/L, manganese 120–180 mg/L and iron 70–80 mg/L. This water is generally mixed with other less polluted water for irrigation in the Jordan Valley area.

Trace elements are also found in the effluents of other industries such as:

Iron: Eagle Distilleries 3.68 mg/L, United Distilleries 4.82 mg/L, Oven Manufacturing 1.82 mg/L.
Nickle: Warehouse Manufacturing 0.32 mg/L, and Hussein Thermal Station 0.21 mg/L.
Zinc: Arab Brewery 0.41 mg/L, Hussein Thermal Station 0.43 mg/L, Hussein Iron & Steel 0.23 mg/L, Arab Co. for Commerce and Food Industries 0.432 mg/L, United Industries 0.31 mg/L, Warehouse Manufacturing Co. 0.532 mg/L, Oven Manufacturing 0.2 mg/L and Jordan Matches 0.72 mg/L.
Lead: National Industrial Co. (tissue paper) 0.045 mg/L.

Organic Loads
The majority of industries (food, drinks, paper, yeast, polymers, detergents … etc.) produce effluents with high organic loads expressed in their BOD_5 contents

Table 5.5 Examples of industries with high organic load in their effluents (Salameh 1996)

Type of industry	BOD_5 mg/L	Flow m³/day
National industries (tissue paper)	156	104
Sulfo-chemicals	805	62
United industries (distillery)	132	35
Arab Co. for commerce and Food Ind.	850	85
Chemical industries Co. (detergents)	613	8
Yeast Co.	8120	330
Eagle distilleries	3212	130
Tanning industry	134	370
Arab brewery	890	45
Jordan hygiene paper	612	105
Jordan matches	1500	4.3
Textile Co.	280	21
Imperial underwear	170	52
Chemical polymers	1090	4.1

(Table 5.5). The BOD_5 values in those effluents exceeded 100 mg/L and reached a value of 8120 mg/L in the Yeast Co. effluent.

The weighted average of BOD_5 for the effluents of industries listed in Table 5.5 is 1961 mg/L which is almost double the value for raw domestic waste water in Jordan.

Organic loads such as those contained in the effluents of the industries listed in Table 5.5 are degradable under natural conditions. But, they add to the pollution of surface and groundwater resources along the wadis where they flow or in dams where they collect. They also enhance eutrophication processes.

High Salinity

High salinity waste water is produced by the industries listed in Table 5.6. Some of these not only produce brackish water with EC values of a few thousand µS/cm,but also saline water with EC values of more than 10,000 µS/cm, such as the Tanning industry with an average EC of 17,650 µS/cm, Battery industry 15,150 µS/cm, Chemical Industries (Detergents) 42,160 µS/cm and Sulfo-chemicals 11,150 µS/cm.

In a country with scarce water resources and generally higher salt contents in its natural water resources, industrial effluents with high salinities have severe conse-quences on its surface and groundwater resources.

For example, to make the effluent of the Tanning industry suitable for irrigation (1500 µS/cm) from a salinity point of view 8500 m³/day of a normal water source with an EC value of 800 µS/cm is needed. This means a dilution of about 1:25.

To make the waste water produced by the industries listed in Table 5.6 suitable for irrigation (1500 µS/cm) by dilution from a normal source (800 µS/cm), 30,000 m³/day, or 11 million cubic meters per year of water is required.

This calculation indicates that the problem of industrial water in Jordan is mainly a salinity problem and to a lesser extent a trace element problem or an organic load problem.

Table 5.6 Example of industries with high salinity concentration in their effluents

Industry	EC value (µS/cm)	Flow (m³/day)
Tanning	17,650	370
Eagle distilleries	2515	130
Yeast Co.	7750	330
Battery	15,150	12
Iron & steel Co.	5130	265
Chemicals industries Co. (detergents)	42,160	8
National industries (tissue paper)	3876	104
Petroleum refinery	3655	4600
Hygiene paper	3240	105
Sulfo-chemicals	11,150	62
Jordan tiles	5325	57

Other Industrial Pollutants

The waste water effluents of some industries contain specific substances which may negatively affect the environment. Examples of these are the phosphates discharged with the effluents of ICA (Intag) with a concentration of 560 mg/L and Warehouse Manufacturing with 62 mg/L, compared to domestic waste water treatment plants' effluents of up to 50 mg/L. Phosphates in water enhance the eutrophication processes in surface water bodies and lead to ageing lakes.

Also the ABS values of ICA (Intag) and Arab Detergents effluents are 720 mg/L and 1250 mg/L compared to effluents of mechanical waste water treatment plants of domestic origin with up to only about 10 mg/L or to stabilization ponds with 5–45 mg/L.

5.4   Pollutants of Emerging Concern in Water and Waste Water

Pharmaceutical residues, biocides, industrial and household chemicals (henceforth; upcoming pollutants) may be added to the environment and to the different types of water sources and waste water through a variety of pathways and mechanisms, and may not be fully eliminated in waste water treatment and thus distributed widely in the aquatic environment.

In the same way that these pollutant compounds harm and kill mosquitoes, flies, plant disease-causing organisms … etc., they also affect and may be fatal to human beings and other living creatures. The difference is the doses which a body of a certain size and weight receives and can tolerate. But most of these chemical compounds are persistent in nature; they need many years, in some cases tens of years, to disintegrate to 80–90% of their original activity.

Therefore, their effects are cumulative. This is to say that taking a small dose (more than a body can assimilate) for a long time causes illness and may be lethal.

Now it is generally recognised that such chemicals are not a safe interference with natural processes but a risky one for human health. If these compounds reach surface and/or groundwater bodies the water becomes contaminated, and depending on their concentrations unsuitable for human consumption or even for fish farming, irrigation or other similar uses.

When these chemicals reach fields, water bodies and other sites, they contaminate soils, waters and native plants. The pathways of upcoming pollutants into water bodies are:

Surface Water
- Waste waters of households and industry, directly or through the effluents of their waste water treatment plants.
- Solid wastes leachates to surface and groundwater bodies.
- Direct application to stagnant or flowing waters to control mosquitoes, flies … etc.
- Direct deposition from the atmosphere or via dust particles, plants.
- Washed from soils and plant remains by runoffs due to excess irrigation and precipitation events.

Groundwater
- Leaching of contaminated soils irrigated with water contaminated with chemical compounds.
- Infiltration of contaminated water along wadi courses or from surface water bodies; reservoirs, waste water treatment plants etc.

5.4.1 Upcoming Pollutants in Treated Waste Water

Treated waste water is an essential water resource for irrigation in Jordan and is considered as an alternative water source to substitute fresh water where it is suitable quality-wise. In Jordan about 85% of the treated waste water is used in irrigation, mostly after being mixed with fresh water. The fact that upcoming pollutants' compounds occurring in municipal waste waters are often only incompletely removed in the course of waste water treatment deserves special concern, as these are introduced to receiving water bodies used for irrigation or are directly used for irrigation. The effluents may also be stored in dams or infiltrate to recharge the groundwater which might later be used for drinking purposes.

Due to the importance and vital role of treated waste water use in irrigation in Jordan it is imperative to evaluate whether the present practice is safe with respect to the introduction of such compounds into the food chain through the soil and the plants grown partially or completely on lands irrigated with treated municipal waste waters.

The studies carried out in Jordan were restricted to Khirbet es Samra treatment plant, Zarqa River and King Abdullah Canal waters, and some groundwater samples

Table 5.7 Average concentrations (ng/L) of some upcoming pollutants in surface, ground and treated waste water in the Jordan Valley area (Zemann et al., 2014)

Source, no. of samples/substance	Ground water 43	Surface water 34	Treated waste water 18
Carbamazepin	74	800	3500
Ibuprofen	56	80	250
Fenofilbrate	74		260
Diatriazoic acid	120	140	120
Iohexol	31	654	290
Iopromide	74	650	860
Iomeprol	29	2200	1400
Iopamidol	59	400	6600
Iotalamic acid	10	21	42

in the Jordan Valley area. These studies were intensive and incorporated analyses of water, soils and plants irrigated with that water (Table 5.7).

Worldwide, more than 100 upcoming pollutant compounds have been detected in surface water bodies receiving treated waste water, a fact which highlights the potential health risk if introduced to the food chain via plants irrigated with that water, drinking water or the meat of animals fed on forage growing on the contaminated water.

In addition to pharmaceutical compounds (PC), industrial and household chemicals and biocides which are not fully eliminated in waste water treatment also become distributed widely in the aquatic environment. Examples are corrosion inhibitors or sulphonamides. Contrary to others, pharmaceutical compounds and biocides are of specific eco-toxicological concern because they are designed to induce biochemical effects.

The occurrence of upcoming pollutants in soils and plants has been comparatively well investigated in the past decade (De Linguoro et al. 2003; Thiele-Bruhn 2003; Kreuzig et al. 2003; Sarmah et al. 2006; Blackwell et al. 2007; Davis et al. 2006; Kinney et al. 2006; Walters et al. 2010; Carr et al. 2011).

The half-lives of contaminants in soil are governed by sorption, biodegradation, the formation of bound residues and leaching and are influenced by the physico-chemical properties of the compounds and the soil's characteristics, e.g., pH and organic matter content (Walters et al. 2010; Scheurer et al. 2012).

As regards the irrigated lands in Jordan, the studies conducted by Riemenschneider et al. (2017) did not detect concentration decreases in the water withdrawn by suction cups from different depths. This suggests that during pulse irrigation of highly permeable soils the residence time of irrigation water is too short to allow for significant biodegradation.

Field studies along the Zarqa River show that some of the 33 analytes (23 chemicals +10 metabolites) were repeatedly found in the studied crops. Also here Carbamazepine (CBZ) and some of its metabolites were most prominent and the CBZ concentration was higher in roots than in leaves.

Sulfamethoxazole, diclofenac and carbamazepine are the most prominent examples of PCs which are not efficiently removed or altered in municipal waste water treatment. Beyond PCs however, also industrial and household chemicals may not be fully eliminated in waste water treatment and thus distributed widely in the aquatic environment if their polarity is high. Examples are corrosion inhibitors such as benzotriazole and tolyltriazole or sulfonamides.

Residues of various Pollutants of Emerging Concern (PECs) have recently been shown to be absorbed by edible plants when the PCs were introduced either by the spiking of the medium or irrigation water, or by sludge application.

Previous studies show that carbamazepine, lamotrigine, caffeine, metoprolol, sulfamethoxazole and sildenafil are persistent in soils when introduced via treated waste water. The findings show that the rapidly degradable compounds were not detected, while others were present throughout the whole of the soil profiles without any concentration decreases in the water withdrawn from different soil depths. This suggests that during the pulse irrigation of highly permeable soils the residence time irrigation is too short to allow for significant biodegradation.

Studies should now be aimed at:

- Improving the scientific understanding of the biodegradability of PEC originating from treated waste water in soils, and during movement to the groundwater and the uptake of PEC originating from treated waste water into edible plants in semi-arid regions.
- Providing mechanistic information about the uptake, translocation and metabolism of PEC that will allow us to predict the behavior of other PEC based on their physicochemical nature.
- Elaborating recommendations for the safe reuse of treated waste water in agriculture.
- In Jordan the analyses show that the concentrations are generally still below the recommended guidelines for the specific use of the analyzed water sources.

But, the presence of these biocides in some surface and groundwater resources indicates that they and other not examined sources may represent a potential danger for the water uses.

5.4.2 Over-exploitation, Resources Depletion and Aquifer Salinization

The information and text of this chapter are based on Salameh (2008) '; numbers and data have been updated according to available information in MWI (open files).If the extracted amount of a groundwater body exceeds the average recharge over a series of years, and if it is practically not expected that under the normal climatic conditions very wet years may compensate for the extracted amounts, the water body is considered to be over-exploited. In this case the groundwater levels or piezometric heads drop monotonously, although with some seasonal or secular fluctuations.

This situation is affecting almost all areas in Jordan; the Jafr, Azraq, Dhuleil, Shoubak, Agib, Qastal, Qatraneh, Wadi Arab, Northern Badia, Amman-Zarqa and Disi areas. In all these areas water levels are continuously declining, and wells drilled in Dhuleil, Azraq and Amman-Zarqa have to be deepened and pumps sunk in accordance with continuously dropping water levels.

Dropping water levels result in the coning up of underlying water bodies which contain generally higher salinities. Hence, their upcoming causes the salinization of the water bodies under use. In the Dhuleil and Badia areas, increasing salinities due to upcoming are affecting the aquifers under exploitation.

Salty water is also found both at the surface of and underlying playas and certain desert oases. Oases areas are generally fed by surface and/or groundwater, where the water evaporates and salts accumulate or remain dissolved in the rest of the water.

Jafr Basin

In the early 1960s the Jafr groundwater basin was developed to provide water for both domestic and irrigation uses. After a few years of water extraction, groundwater levels started to drop rapidly and the water quality began to deteriorate. Therefore, part of the investment in that area was lost because of the worsening water quality which made it unsuitable, even for irrigational use. Water levels also dropped to uneconomic levels.

Table 5.8 shows a comparison of the water quality in well No. 17 in that area and illustrates the gradual salinization of the aquifer's water.

Dhuleil Area

The water resources of the Dhuleil area were developed in the early 1970s. At that time the water quality of the aquifer was excellent for different uses. An increasing number of wells were licensed and drilled in rapid succession. The water found its use in agriculture. By the end of the 1970s the increasing salinities of the aquifer water forced some farmers to excessively irrigate their fields. Gradually, in the 1980s more and more land was laid fallow. The results of this development were the salinization of the aquifer due to over-pumping and of the soils due to the use of brackish water and fertilizers, and as a result, further salinization of the aquifer's water due to irrigation return flows.

Table 5.9 shows the development of the groundwater quality on the example of well Al 1109, Wasfi Al- Tal

The salinization of the aquifer's water can be deduced from the rapid increase of Na, Cl, SO_4, Ca and Mg concentrations. Irrigation return flows were indicated by the increase in the nitrate value from 15 mg/L in 1970 to 120 mg/L in 1995. Also these wells showed high concentrations of phosphates in 1995, which can be attributed to fertilizers used in agriculture. Rimawi (1985) studied the groundwater resources of the area and concluded that both over-exploitation and irrigation return flows were leading to the rapid deterioration of the water quality.

The drop in water levels is illustrated on the examples of wells No. 5, Al 2698 (Hallabat observation well 1) and Al 1040, TW 5. The average declines in the water table from 1980–1988 in well Al 1040 was around 25 cm/year. After that it increased to around 85 cm/year in both wells.

Table 5.8 Salinity increase in Jafr well No. 17, mg/L (Files of WAJ)

Date	1965	1973	1994	2009
EC	909	3.100	3760	3890
pH	7.5	7.2	7.58	7.76
Ca^2	82	164	233.6	245.2
Mg^2	40	100	122.2	145
Na^+	85	287	364.4	386
K^+	7.5	16	20.0	22
Cl^-	99	768	933	990
SO_4^{-2}	145	240	333	320
NO_3^-	10	19	11.8	12.0
HCO_3^-	321	256	217	298

Table 5.9 Water quality development of well no. Al 1109 from 1970–1995 (in mg/L)

Date	1970	1979	1995
EC	440	1060	2700
pH	8.1	7.8	
Ca^2	12	51	159.3
Mg^2	16	42	115.5
Na^+	53	103	180.7
K^+	4.7	9.0	18
Cl^-	61	272	695
SO_4^{-2}	29	71	126.3
NO_3^-	128	82	72.6
HCO_3^-	15	23	120

5.4.3 Water Levels

Azraq Area

Since 1982, water produced from a series of wells located to the north of Azraq has been pumped to Amman to alleviate increasing demand. The amount of water pumped to Amman ranged from 12–16 MCM/year. By 1986 the drawdown in the groundwater table reached 3 m.

Well drilling activities in the Azraq area intensified in the late 1980s and in the first few years of the 1990s. The results of groundwater salinization and the drop in water levels extended to affect the whole basin.

At present the general drawdown is 38 m. The spring discharges in South Azraq, and those of Qaisiya and Soda springs stopped in 1993. The total groundwater natural discharge of 16 MCM/year, measured by Arsalan in 1973, decreased because of over-pumping to 10 MCM/year in 1983, and to zero in the past 4 years.

The drop in water level is illustrated on the examples of wells F1043 (Azraq) and F 1280 (observation AWSA-2) with an average decline in the water table of 100 cm/year (Fig. 5.3).

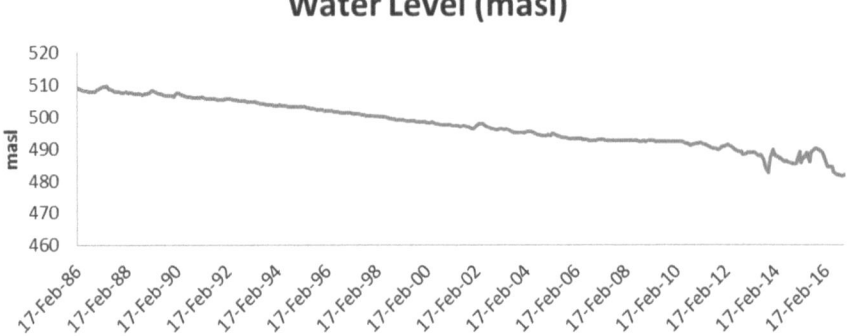

In 1994/1995 a decision was taken to pump around 0.5 MCM of fresh water produced from the Water Authority wells (AWSA wells) into the oasis which fell dry in 1993. This action was initiated by a Global Environmental Facility (GEF)-UNDP project, Rasmar, to somehow restore the oasis functions with the assistance of MWI. But, the impacts of such an action on the groundwater resources may be highly detrimental in light of the hydrogeological situation prevailing in the area.

The oasis is an exitless evaporation pan. Before the developments of the last 15 years the water system of the oasis functioned in the following pattern:

- Floodwater originating within the catchment area collected in the marsh land area of a few square kilometers.
- The shallow aquifer B4, the Basalt or the composite aquifer of B4-Basalt and recent sediments discharged their groundwater directly into the oasis pools through a few springs such as Ora, Mustadima, Soda, Qaisiya and others.
- The groundwater in the deeper aquifer, the B2/A7, is pressurized with a piezo-metric head exceeding that of the shallow aquifers basalt-B4 recent deposits and that of the deep Kurnub aquifer. Therefore it leaks upwards through the overlying B3 into the shallow aquifers and from there into the oasis, and downwards through a series of aquifuges into the Kurnub.

All the water which used to flow into the oasis and marshes evaporated and left the salts behind.

Gradually, the increasing pumping out of the aquifer has led to the following:

- Floodwater is still reaching the marsh land although in decreasing quantities.
- The shallow aquifer water levels have dropped and springs have stopped feeding the oasis.

The deeper groundwater still leaks upwards into the shallow aquifers but its water is not appearing at the surface (Fig. 5.4).

The marsh lands surrounding the Azraq pools are built up of silt and clay with gypsum and halite. The salt concentration in the ditches dug in the area to extract

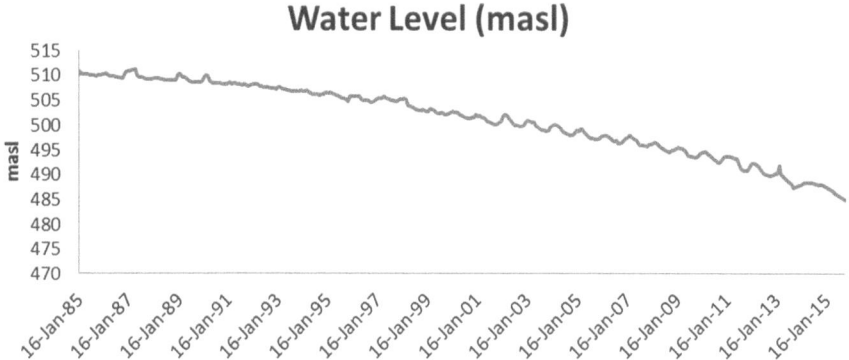

Fig. 5.4 Water level development of well no. F1280 (AWSA Observation 2)

table salt reach sometimes 70 g/L of water. These marshes extend to the pool areas and border them.

But, since the water table in the area has dropped by about 30 m (1982–2016), the areas underlying the pools and marshes have turned into unsaturated zones. Filling the pools with water will cause the water to infiltrate to the groundwater table taking with it the dissolved salts thereby increasing the groundwater salinity. Now since the pools will function as a recharge mouth, the water levels beneath them will rise and the groundwater flow direction will be reversed towards the production wells. This will result in increasing salinities of production wells in the surroundings of the oasis including the wells of AWSA.

Therefore, it is expected that the Azraq project (UNDP/Rasmar) will cause an increase in the groundwater salinity, and if it continues for the coming few years some well waters will become unsuitable for the domestic water supply.

It is somewhat surprising that the GEF through the UNDP-Azraq project (Rasmar) is undertaking an action which will certainly lead to salinization processes in the vital groundwater resources of the country.

Disi-Mudawwara-Sahl-es-Suwwan Area
The sandstone aquifer in this area contains water with a recharge age of 2000–1100 years; hence it can be classified as fossil water. Major development in the unconfined parts of the aquifer (Disi) and the confined parts (Mudawwara) started in 1985–1987. From that time on, increasing groundwater amounts have been pumped out of the aquifer. The extraction for the water supply of Aqaba town and for the local irrigation projects until 2013 (start of water pumping to the northern areas of Jordan) amounts to more than 85 MCM/year.

From 1985–1987, water levels in the area started to drop (Fig. 5.5). The average drop in level is around 0.6 m/year in Khreim and Mneisheer. This drop in groundwater levels coupled with the very scarce rainfall in the area with an average of 30 mm/year indicates that the aquifer is undergoing a mining process, which means that the aquifer is being emptied; over-exploited.

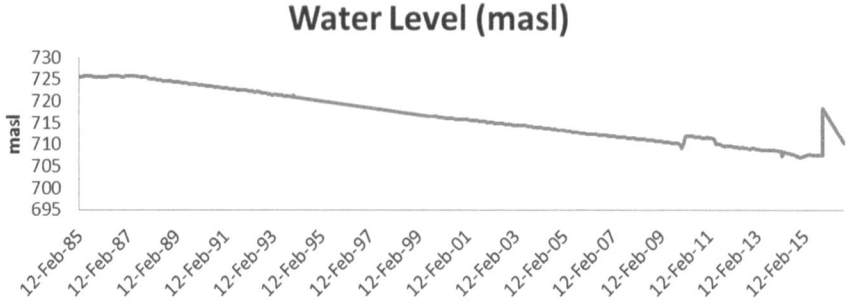

Fig. 5.5 Water level development of well no. ED 1202 (Mneisheer observation 2)

Until now only slight changes in the water quality have taken place, but irrigation return flows will certainly reach the groundwater in the unconfined portions of the aquifer (Disi) which may lead to increasing salinities and increasing fertilizer concentrations such as nitrates, phosphates, potassium etc.

In the confined portions of the aquifer the salt water found in the Khreim confining layer may, upon pressure releases due to water level declines, lead to salt releases from the Khreim unit into the underlying Disi aquifer. Some signs of this are seen in Mudawwara wells.

The North-Western Parts of the Highlands (Sama Sirhan, Mafraq, Wadi Arab, and Mukheiba)
The main aquifer in this area is the B2/A7. Extraction of water from the unconfined eastern part of the aquifer (east of Irbid to Mafraq) was very limited until the beginning of the 1980s. After that intensive well-drilling activities were carried out and hundreds of new wells were put into production to supply water for irrigation. Wells were also drilled in the confined, western part of the aquifer (Mukheiba, Wadi el Arab, Waqqas, Abu Ziad and others). The results of the extensive water extraction were dropping water levels and a declining water table. The parts of the confined aquifer west of Irbid became free water table aquifers. Figures show the drawdown in water levels in the area. On average for the last 8–20 years the declines in water levels for the different wells ranged from 0.55 cm/year in Ilyan Migbel well in the unconfined part of the aquifer to 3.2 m/year in Kufr Assad well in the confined part of it (Fig. 5.6).

No major changes in the composition of the groundwater were registered in the western confined and unconfined parts of the aquifer; Wadi Arab wells, Mukheiba wells, Nuayma and Yarmouk University wells.

In the eastern parts, irrigation return flows seem to affect the chemistry of the water, but there are no signs of salinization due to over-pumping.

Qatraneh Area
Qastal, Siwaqa Qatraneh Sultani and Hasa are understood as comprising this area. The number of wells in the area is around 250 including some 70 governmental wells. Governmental wells in Sultani supply Karak City with municipal water,

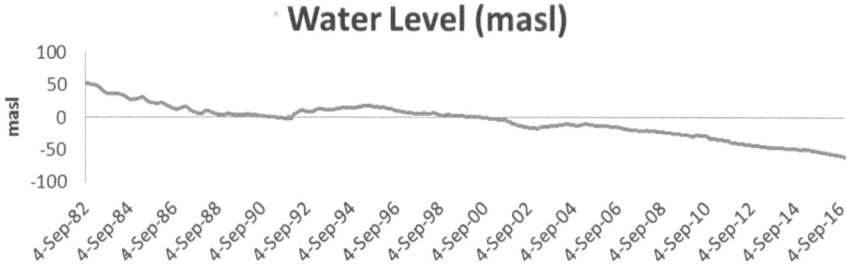

Fig. 5.6 Water level development of well no. AE 1003 (Kufr Assad)

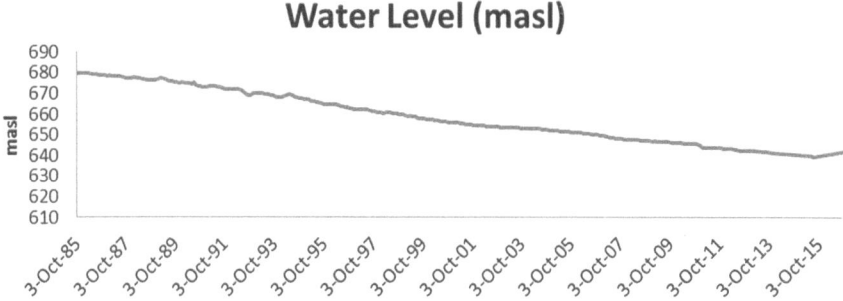

Fig. 5.7 Water level development of well no. CD 1106 (Qatraneh observation 10)

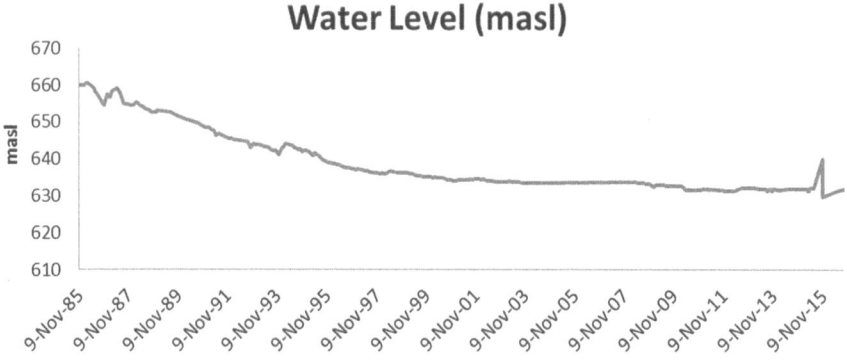

Fig. 5.8 Water level development of well no. CD 1132 (Siwaqa Observation 2)

whereas Qastal, Siwaqa and Qatraneh are connected to the water supply of greater Amman.

Major drilling and extraction activity started in 1984/85. At present around 55 MCM/year are extracted from the B2/A7 aquifer extending from Hasa to Qastal. The result of this extraction can be seen from Figs. 5.7 and 5.8 which illustrate the drawdown in observation wells Qatranah no. 10 and Siwaqa no. 2. The drop in water levels in these wells ranged from 1.27 and 1.1 m/year.

As regards the chemical composition of the groundwater no major changes seem to have taken place in the last decade although water levels are rapidly dropping. The reason for that could be the presence of a thick aquiclude underlying the B2/A7 aquifer from the saltier deep groundwater bodies.

Only in a few wells have sudden rapid increases in salinities taken place. But these increases are temporary and the salinity drops again after aquifer releases. It seems that this phenomenon is restricted to wells close to faults or fault zones where drawdowns in wells due to pumping cause the temporary upcoming of saltier water along these faults.

5.4.4   Wasted Groundwater Resources and Their Impacts

In the 1990s the Jordan Valley Authority drilled numerous wells along the Jordan part of the Rift Valley, particularly in the area close to the foothills (Hisban, Kafrain, Rama, North-Shuna, Ghor Haditha and other places). These wells produce artesian, salty thermal water from a variety of aquifers (Kurnub, Zarqa, B2/A7, B4 and combinations of them) covered by confining rocks (B3, B1, A5, 6, A1, 2, 3 and recent marly deposits).

The high salinity of these wells did not allow any relevant use. Also, because they are artesian they continued to flow (to discharge water) during the early to mid-1980s without the appropriate efforts of the responsible agency to close these wells and stop their wasted discharges (no use) in the interest of the country's water supply.

Concerning that, two facts are worth mentioning here:

1. Although the water flowing from these wells is salty, the water bodies from which they produce extend in an easterly direction and there contain fresh water, e.g., B2/A7 of north Shuna corresponds to the B2/A7 of Wadi Arab wells and to the municipal and irrigation wells east of Irbid, Nuayma, Yarmouk University … etc.; Rama wells produce from the same Kurnub-Zarqa aquifer complex producing fresh water in Baqa'a and west of Mahis.
2. The eastern extensions of the water-producing aquifers and the water contained in them build the backbone of the overlying fresh water bodies which form irreplaceable sources for the water supply of the country.

The conclusion is now that the salty water flowing out of the wells in the Jordan Valley area belongs to the same water bodies producing fresh water in the highlands. Therefore, not stopping the useless discharges of these wells will have the result of partly emptying the aquifers or draining the water bodies with the following consequences:

1. The fresh water in the eastern extension of the water bodies will move faster westward, down-the-gradient to the well's areas where the aquifer matrices are salty causing the salinity of the fresh water to strongly rise.

2. Generally, the eastern extensions of the water bodies from which the above mentioned wells produce build the basis water body over which all shallower groundwater bodies rest in a hydrodynamic equilibrium. The draining of these underlying groundwater bodies means that the overlying ones will have to substitute any water deficit arising in the lower groundwater bodies. This leads to declining water levels and salinization of the fresh water introduced into the salt water confined aquifer.

Allowing the wells to flow uncontrolled and the water to go unused, as it has for the last 10–15 years is depriving the country of vital groundwater resources and undermining its water supply system.

Such actions and projects leading to the depletion and salinization of water resources without any benefit to anybody, whether intended or unintended, can be considered as an "act of terror" against the interests of the nation and its resources base. Hence, they should be stopped immediately.

References

Batarseh MI, Kreuzig R, Bahadir M (2003) Residue analysis of organic pollutants in sediments from the Amman/Zarqa area in Jordan. Part I: development of analytical methods and distribution patterns of PAHS. Fresenius Environ Bull 12(9):972–978

Blackwell PA, Kay P, Boxall ABA (2007) The dissipation and transport of veterinary antibiotics in a sandy loam soil. Chemosphere 67(2):292–299

Carr DL et al (2011) Microbially mediated degradation of common pharmaceuticals and personal care products in soil under aerobic and reduced oxygen conditions. Water Air Soil Pollut 216(1-4):633–642

Davis AP et al (2006) Water quality improvement through bioretention media: nitrogen and phosphorus removal. Water Environ Res 78(3):284–293

De Liguoro M et al (2003) Use of oxytetracycline and tylosin in intensive calf farming: evaluation of transfer to manure and soil. Chemosphere 52(1):203–212

El-Natour RM (1993) Solid waste management in Jordan. WHO. (in press)

Farquhar GF (1989) Landfill leachates migration in soils. Can Geotech J 6:51–87

Kinney CA et al (2006) Presence and distribution of wastewater-derived pharmaceuticals in soil irrigated with reclaimed water. Environ Toxicol Chem 25(2):317–326

MWI (Ministry of Water and Irrigation) Jordan. Open files (2016)

Riemenschneider C, Seiwert B, Goldstein M, Al-Raggad M, Salameh E, Chefetz B, Reemtsma T (2017) An LC-MS/MS method for the determination of 28 polar environmental contaminants and metabolites in vegetables irrigated with treated municipal wastewater. Anal Methods 9(8):1273–1281

Rimawi O (1985) Hydrogeochemistry and isotope hydrology of the ground and surface water in North Jordan. Ph.D. thesis, TUM, München

Salameh E (1996) Water quality degradation in Jordan. Friedrich Ebert Stiftung, Amman and Royal Society for the Conservation of Nature, Amman, 179 p

Salameh E (2008) Over-exploitation of groundwater resources and their environmental and socio-economic implications: the caseof Jordan. Water Int 33(1):55–68

Sarmah AK, Meyer MT, Boxall ABA (2006) A global perspective on the use, sales, exposure pathways, occurrence, fate and effects of veterinary antibiotics (VAs) in the environment. Chemosphere 65(5):725–759

Thiele-Bruhn S (2003) Pharmaceutical antibiotic compounds in soils–a review. J Plant Nutr Soil
    Sci 166(2):145–167
Walters E, McClellan K, Halden RU (2010) Occurrence and loss over three years of 72 pharma-
    ceuticals and personal care products from biosolids–soil mixtures in outdoor mesocosms.
    Water Res 44(20):6011–6020
Zemann M, Wolf L et al (2014) Sources and processes affecting sptio-temporal distribution of
    pharmaceuticals and X-ray contrast media in the water resources of the Lower Jordan Valley.
    Sci Total Environ 488–489:100–114

Chapter 6
Water Pollution Management and Cost

6.1 Management and Cost

The efficiency of water pollution management can be expressed by its impacts on the environment. Efficient water use, waste water treatment and reuse and the avoidance of negative environmental impacts are good indicators of sound management. The misuse of water resources, water and environmental pollution, over-exploitation of water resources, violating sustainability principles and not respecting intergeneration equity indicate that water resources management is inefficient. Using social and economic development to justify water and environmental pollution cannot be accepted whether the pollution and depletion are quantitative or qualitative. Because by doing that the development runs into a vicious circle, and the damage to the environment and water resources reflects negatively on the development itself.

Therefore, development should be in line with the environmental and water resources ensuring their sustainability. This means that the management of water and environmental resources must be sustainable and secure to ensure the stability and the rights of future generations in respect of water resources. Therefore, the management of water resources must reflect sound economics. And as a result, projects generating the degradation of environmental or water resources without sound economic plans and instruments to repair the degradation can be considered as projects of misallocation and misuse.

This makes it conditional that the development of water resources for whatever use should include environmental, technical and economic feasibility studies with the objective of being beneficial to the society. Not fully including water and environmental issues in the feasibility studies of water resources development, waste water treatment and reuse means the benefits of the development to the society will be temporary and in fact detrimental to the society in the long-run.

The text of this chapter is based on Salameh (1996) and Salameh (2008) with updated data from MWI (2016).

© Springer International Publishing AG, part of Springer Nature 2018   111
E. Salameh et al., *Water Resources of Jordan*, World Water Resources 1,
https://doi.org/10.1007/978-3-319-77748-1_6

The objectives of water resources development should aim at serving the society with the main target of conserving the environment and protecting the water resources for long-term sustainable yields to serve future generations. These objectives can be summarized as follows:

- Developing water resources in a sustainable way to serve present and future generations
- Covering the basic needs of the population by providing clean and sufficient amounts of water to guarantee health and hygiene
- Ensuring a healthy environment in all aspects of water development such as resource development, conveyance, use, treatment and reuse of waste water.

Therefore, the development of water resources should be preceded by studies on the impacts on the water resources themselves in terms of quantity and quality, on the environment, on future generations' rights to the resources and on their long-term social and economic soundness.

Our concern now is how to address water degradation and depletion issues in economic terms in order to achieve environmentally relevant water management conditioned by intergeneration equity and economic feasibility.

The main objective of evaluating water resources pollution and damage in economic terms is to facilitate a better understanding of the human impacts on these resources. The value assigned to a water source depends on several factors such as its use, its relative scarcity, the demand for it, and its socio-economic relevance. Since assessing the economic value of water resources development incorporates a socio-economic component it becomes difficult to assign the benefits and costs of water use for the society, the environment, the health of man and other living organisms, the economy of individuals and governments and politics.

The degradation of water resources, whether quantitatively or qualitatively, can also be approached economically by evaluating the economic losses due to the degradation or complete loss of water resources. In this case one can approach the evaluation of water degradation by two means:

- Calculating the cost of degradation abatement; the cost incurred to reduce or eliminate the degradation causes, e.g., adequately treat pollution at source.
- Calculating the cost of damage due to degradation of water resources (e.g., decreasing productivity of irrigation water due to increasing water salinity).
- Pollution and degradation reduce the value of the affected resources, whether water, soil, crops ... etc., which are common goods of the society. For, cost is involved in restoring the affected environmental elements.

Sometimes it is essential to compare both costs of 1 and 2 above in order to illustrate the severity of the problem once degradation has spread from a limited place in defined amounts into larger areas thus increasing the extent of damage'.

If practically, the cause of degradation can be eliminated at a certain cost, then not paying this cost can be considered as the cause of degradation.

The principle that the polluters should bear the full cost of the repair of any water resources or environmental damage for the welfare of the society indicates that water resources and environmental degradation are economic problems.

Not included in the cost mentioned above is the social cost, that is, the detrimental consequences of water resources degradation for the physical and mental well-being of members of the society (e.g., social cost of odorous waste water treatment plants). Therefore any valuation of the degradation of resources can only be sketchy, because the damage may involve significant psychological elements, such as loss of pleasure owing to the pollution of a surface water body, for example a lake (no swimming, no fishing, bad smell), or visual pollution or pain and suffering.

6.2 Examples of Degradation Cost

6.2.1 Waste Water Treatment

As mentioned in the chapter about pollution, waste water treatment in Jordan is undertaken by a variety of methods with different effluent qualities and different costs. Table 6.1 shows the average operation cost of treatment for some domestic waste water plants in Jordan for the year 2015.

This table indicates the following facts:

- The treatment in stabilization ponds (S) is cheaper than in mechanical treatment plants (M). It is only about half the cost of treatment in mechanical plants.
- Within stabilization ponds the treatment cost ranges from around 14 fils/m^3 in Khirbet-es-Samra up to 55 fils/m^3 in Mafraq WWTP.

Table 6.1 Waste water treatment operation cost (fils/m^3)

Treatment plant	Average cost	Type of plant	Status
Khirbt es-Samra	14	S	Replaced
Aqaba	42	S	In operation
Madaba	45	S	Replaced
Mafraq	55	S	In operation
Ramtha	54	S	Replaced
Ma'an	42	S	Replaced
Irbid	150	M	In operation
Jarash	120	M	In operation
Kufranja	145	M	In operation
Baqa'a	134	M	In operation
Abu Nuseir	195	M	In operation
Salt	104	M	In operation
+ Karak	165	S + M	In operation
+ Tafilah	168	S + M	In operation

- The cheapest cost per m³ in a mechanical treatment plant is 104 fils in Salt and the highest cost is in Abu Nuseir with 195 fils.

At first glance, stabilization ponds seem to be a very attractive option for waste water treatment from a cost point of view. But, if the effluent qualities of these stabilization ponds and their environmental impacts are considered the whole picture may change. The impacts of these WWTP have numerous economic dimensions such as:

- High evaporation losses.
- Degrading qualities of groundwater bodies underlying the treatment plant and along wadis where effluents are discharged.
- Deterioration of surface water qualities along wadis and river courses, in surface water reservoirs and finally in areas where the water is reused in irrigation.
- Health detriments to the population living in the surroundings of these plants and along the effluent courses.
- Deterioration of the quality of life for the affected population groups.
- Health impacts on the livestock drinking the effluents or feeding on plants irrigated by these effluents.

Of the above mentioned impacts only the first three can be adequately approached in economic terms.

The analysis of waste water treatment projects in Jordan lead to the following findings:

- All stabilization ponds produce an effluent with high BOD_5 concentration exceeding the Jordanian standards for waste water treatment effluents.
- All mechanical treatment plants produce an effluent with a BOD_5 in accordance with the Jordanian standards for waste water treatment effluents.
- The cheaper treatment in stabilization ponds is strongly reflected in the bad effluent quality.

For retention times ranging from 28 to 40 days and evaporation rates ranging from 2800 mm/year in Ramtha to 4000 mm/year in Aqaba evaporation losses in stabilization ponds range from 18% to 38% of the inflowing amounts into these treatment plants compared to 2–3% in mechanical treatment plants.

The weighted average evaporation losses of all stabilization ponds amount to 25% of the incoming water, which means a loss of water resources of 22 MCM/year with a value of 3 million JD. But, if the productivity of the water in irrigation is considered (2 JD/m³) the loss for the nation equals 44 million JD/year. The other result of evaporation loss is the salinity concentration in the remaining water which equals, in the case at hand, a 30% increase over the original salt concentration of around 1500 μS/cm.

For the different stabilization ponds the effluent salt concentration is 22–70% more than the inflow concentration.

Such an increase means further deterioration of water qualities. The reduction in the productivity of such water when used for irrigation compared to its productivity at normal salinity ranges between 10% and 30% for the main crops produced in Jordan. The average productivity reduction is around 20%. Even if the water is diluted by better quality water in a ratio of 1:2 (better quality to treated waste water) the result will still be a reduction in crop productivity of around 15% of the original productivity of the mixed water. But because effluents are not always diluted and dilution water is not always available, it is estimated that losses are in the range of 18% of the original productivity for the main crops in Jordan.

If the average productivity of 2 JD/m³ of water is reduced by 18%, the average productivity losses of the 65 MCM/year of stabilization ponds effluents amount to 23.4 million JD per year.

In addition, the waste water of stabilization ponds infiltrates through their unsealed bottoms to the groundwater bodies. Also the effluents flowing along the downstream wadis infiltrate and recharge the groundwater. Because of their high organic and salt contents, and due to their ability to react with rocks (aggressively), infiltrating waste waters and effluents become saltier, and thus cause rapid increases in groundwater salinities and organic matter contents (Wadi Dhuleil, the Hashimiya area, Aqaba, Ramtha, Madaba and Mafraq).

The amount of deteriorated groundwater in those areas is 9–11 MCM/year, which means losses of their productivity or of the water bodies themselves.

Many wells and springs along Wadi Dhuleil, Wadi Ramtha and generally down-gradient areas of stabilization ponds (see chapter on pollution) have suffered severe quality deterioration due to pollution and salinity increases. Many of the wells and springs have been abandoned. New projects and new water resources have had to be planned and developed which has meant large investments for the public and private sectors. The estimated operation cost of the new projects is 1.5–2 million JD/year with a capital cost of 15–20 million JD.

In addition to the physical losses of water due to evaporation, to quality deterioration and to productivity losses, the incurred costs to the society (the social cost) such as the detrimental effects on the physical and mental well-being of members of the society caused by, for example, odor, loss of pleasure, or visual pollution have to be taken into consideration when calculating the economic impacts. Although it is very difficult to quantify these losses in monetary terms, they should nonetheless be kept in mind.

Comparing the degradation abatement cost of 1.4 million JD/year to the actual damage of 24 million JD/year or to productivity detriments of 34 million JD/year shows that stabilization ponds are damaging to the water resources of the country and its economy.

Over the last 10 years, since the establishment of the first stabilization pond treatment in Jordan in Khirbet es Samra, this type of treatment has proved to be an environmental and economic disgrace to the country and its people.

6.2.2   Cost of Aquifer Over-Exploitation and Depletion

Groundwater from all the major aquifers in Jordan is extracted for the municipal water supply; irrigation water is produced from most aquifers and in some defined areas, groundwater is extracted for industrial uses.

The main areas producing groundwater for municipal uses are:

- Disi for northern Jordan and Aqaba city
- Azraq for Azraq, Zarqa and Amman
- Siwaqa-Sultani-Qatraneh for southern Amman, Amman, Madaba, Karak, Qatraneh … etc.
- Amman-Zarqa for Amman-Zarqa areas
- Za'atari-Mafraq for Mafraq governorate
- Yarmouk and Wadi El Arab groundwater basin for Irbid governorate and Ramtha
- Shoubak for Shoubak
- Jafr for Ma'an
- Jordan Valley aquifers for Jordan Valley domestic supply.

In the section on pollution it was indicated that all shallow aquifers underlying the highlands are being subject tover-exploitation leading to their depletion and degrading qualities as can be deduced from their salinization.

In Jordan, water in aquifers has been developed to the utmost limits and even beyond so that they are now over-exploited and there are no more aquifers to be developed. Therefore, no reserve aquifers to substitute any of the developed groundwater resources are to be found in the country. This implies that Jordan will not be able to continue supplying municipal water in the same quantities and qualities if any groundwater aquifer in use now depletes or suffers quality deterioration.

From this it can be concluded that for the country's municipal water supply two options remain open:

1. Cutting agriculture in the highlands to stop groundwater depletion and salinization and allow aquifers to recover.
2. Desalination of sea water or import of water from water-rich countries.

Even the supply to cover the immediate needs of the country's growing population can never be met from the present sources unless one or both of the above mentioned options are implemented.

Cutting agriculture does not imply that the saved amounts of water can be allocated for domestic supply because the aquifers are now over-exploited, and the total annual depletion exceeds the safe yields by some 300 MCM/year. This is to say that irrigational water use in the highlands should be cut by 300 MCM/year, in order that no further drop in water levels takes place.

Another fact is that around 335 MCM/year out of around 429 MCM/year used for domestic and industrial water supplies are taken from the highland aquifers.

This amount of 335 MCM/year exceeds by far the renewable groundwater amounts in Jordan of some 275 MCM/year.

As a consequence of that, the fact remains that even if all irrigated agriculture in the highlands is immediately stopped just to arrive at an equilibrium between extractions for domestic and industrial supplies on one hand and renewable groundwater amounts (safe yield concept of aquifers) on the other, this will still not allow aquifers to recover.

If water use priorities are listed, then supplying household water seems to have primary importance because it has irreplaceable benefits for the society in general and for each individual who consumes this water. Making a certain minimum level of water supply available to the population is a merit and hence it is not surprising that the government of Jordan subsidizes water and that the water issue receives much political attention. Such priority cannot be assigned to the other main use sectors; industry and irrigation. Although prioritization here is also somehow necessary, the fact that industrial water uses amount to only 5% and irrigation uses to around 60% of the total uses, and also the fact that water productivity in the industrial sector (m^3/JD output) is 20–30 times the productivity in irrigated agriculture indicates that competition between the two need not be feared in the long-run nation-wide. On a local scale (industrial plant versus irrigation) prioritization may turn to out to be an economic problem; who can pay more or pay the other user? Or industry uses the water first; treat it properly to be reused in irrigation.

The growing population requires additional amounts of domestic water which implies that even if irrigated agriculture with fresh groundwater in the highlands were immediately stopped, increasing domestic and industrial water supplies would not come from the present groundwater resources, if safe yield and supply continuity principles are to be respected.

The above discussion shows that the option of curtailing or even stopping irrigated agriculture using freshwater in the highlands does not mean saving water for other uses, but decelerating aquifer depletion. Hence, the only option for Jordan to sustain and increase its municipal water supply would be to desalinate sea water, or to import water from water-rich countries.

The desalination cost amounts to $US 0.75–1.3 per cubic meter at sea shore. Transportation and distribution costs depend on the supplied area. To supply Amman, for example, with water desalinated at Aqaba, transportation and distribution costs may amount to $US $1.0/m^3$.

This indicates that the option of making desalinated water available in Amman would cost somewhere between $US 1.75 and $2.30/m^3$. In the following, it is considered as $US $2.00/m^3$.

At present, making water available at Amman from a variety of groundwater sources costs around $US $0.33/m^3$. But, Disi water costs around $US $1.30/m^3$.

This shows that each cubic meter of groundwater in the highlands is worth an average of $US $2.00/m^3$.

Hence, assigning an adequate price to the groundwater resources to include them in the process of economic development means that each cubic meter of groundwater in the highlands has a present value of $US 2.0.

Therefore, and because no options are available for additional municipal water supply of the major cities in Jordan other than desalination, the over-exploitation of the groundwater resources of the highlands by 300 MCM/year for irrigation means depriving the country (present and future generations) of $US 600 million/year cost of substituting that water through the only viable solution; desalination, which by far exceeds the total productivity of water in the irrigated sector.

6.2.3   Discussion on Regulation and Scarcity Price of Water Resources and Water Quality Deterioration

For sustainable water supplies yielding appropriate amounts of water with suitable qualities, it is pre-conditional that government interventions are introduced in the form of setting regulations, advancing environmental laws, and defining pollution control standards. Such interventions will ensure the fair assessment of pollution causes and extents and the tolerability of environmental elements to that pollution that can also be accepted by the society according to the standards set. In this case the legal principle that the government can restrict an activity if it is presumed (not proven) to be harmful must be accepted by the society.

Therefore, the polluters must obtain from the government a "permission to pollute," that is, to pollute to a certain degree that is environmentally tolerable according to pre-set standards. The permission should specify the quantity and concentration of pollutants allowed to be released in the environment. Carrying capacities of the environmental elements as the limiting factors of pollutants takes into consideration effects on downstream areas and users. Failing to meet the conditions laid down in In the "permission to pollute," conditions should be specified and exceeding the set limits or polluting without permission should be considered as a criminal offence against the society.

As has been described in the case of waste water treatment in stabilization ponds, the abatement of pollution at source is the least expensive among all pollution reduction costs and hence it should be always targeted as a first choice option for pollution minimization. Therefore, pollution costs should be incurred by the polluters themselves at the place of pollution, by a company, firm … etc., or by the taxpayer in the case of municipal waste water treatment and safe disposal of solid-wastes. But, in order to avoid pollution, polluters should be instructed about the environmental damage (both physical and social) which their pollution may cause compared to the cost of abatement of pollution at source. It should be made clear to them that it is in the interest of the nation that pollution sources are treated at source before they inflate. This requires the development of market prices for the water.

In general, water polluters do not pollute in order to (intentionally or otherwise) harm or to impose the cost on others, but only to avoid the cost of stopping the pollution at its source.

Companies pursue the objectives of increasing their profits and maximizing their revenues by reducing their production costs, but controlling and treating their waste water incurs further production costs. Therefore, they try by all means to disregard any pollution control instructions, especially if the penalties for pollution are less than the cost of abating it at the source.

Unless pollution is controlled and abated at source, pollution abatement costs can be considered as an additional benefit or revenue for the polluter.

In order to sustain the scarce water resources of the country and to conserve their qualities by avoiding over-exploitation and water quality deterioration, the overall effects on the water itself and on the produced goods in industry or agriculture will be price increases. However, the major impact of under-valuing the country's scarce water resources is that goods produced remain under-priced compared to goods produced with economic consideration of water use and its environmental impacts. Therefore, market prices should not only reflect the scarcity of water but also its uselessness resulting from quality degradation.

Water quality degradation and over-exploitation are not only undermining the current welfare status of the material base of the nation, but will also affect the resources base of future generations. Hence water quality degradation and over-exploitation under general water scarcity conditions should urge the responsible authorities to regard water as a national security issue. For each individual of us it should be considered a security problem.

If intergeneration equities are to be addressed, the aggravation of the water scarcity by social and institutional factors (such as inequitable extraction rights, antisocial water resources distribution, ineffective pricing policies, inadequate effluent pollution charges and in general inefficient water resources management) should be also regarded and dealt with as a criminal offence against present and future generations. Therefore, water protection measures in the context of environmental and economic soundness and intergeneration equity should be regarded as a means of achieving the wider objectives of sustainable economic and social growth.

It must also be stated that the willingness and ability of the government to take unpopular measures will largely determine whether the depletion and degradation of water resources will continue to threaten and damage the socio-economic growth and well-being of the population. A substantial commitment from the government is required to make decisions that may adversely affect powerful interests.

In addition to the impacts of water pollution and depletion described above, there may be other far-reaching impacts that are only partly understood at this time. Levels of pollution have been judged according to our present knowledge of their effects mainly on mankind, and not on other species that share the same environmental source. Given the state of our knowledge of these effects, it would be wiseto consider any level of pollution or depletion as a potential problem.

References

MWI (Ministry of Water and Irrigation) Jordan. Open files (2016)

Salameh E (1996) Water quality degradation in Jordan. Friedrich Ebert Stiftung, Amman and Royal Society for the Conservation of Nature, Amman, 179 p

Salameh E (2008) Over-exploitation of groundwater resources and their environmental and socio-economic implications: the caseof Jordan. Water Int 33(1):55–68

Chapter 7
Water Politics

7.1 National Interest

It is in the overall national interest of the country to achieve/aspire to "sustainable water resources development" which according to the World Commission on Environment and Development (1987) is defined as "the ability to meet the needs of the here and now without compromising the ability of future generations to meet their own needs." Figure 7.1 illustrates the different degrees of national interest in relation to the water situation expressing the poverty or richness of a country in terms of water resources. The effects of population growth and living standards on the water situation are also illustrated.

The national interest in water resources in Jordan stems from factors which affect the whole situation and existence of this water scarce country. It is a function of security, stability and sustainability (Fig. 7.2).

The national interest in the water resources should be expressed in general water policy principles which themselves should lead to developing a dynamic water strategy for the country with different scenarios, action plans, programs and projects as illustrated in Fig. 7.3.

For Jordan, sustainable water resources development means to:

1. Guarantee a continuous supply of water to the population of Jordan that is sufficient in quantity and safe in quality. And supply the economic sector with the necessary water after optimizing this need and pricing the water at its real cost; capital, operational and the cost to the nation arising from the consumption of its assets.
2. Protect the national water resources in all aspects, maximize their productivity potentials and optimize their uses.
3. Guarantee next generations' rights whatever the present generations' interests in water exploitation and utilization are. This incorporates the sustainability of resources yield and the conservation of biodiversity.

The text of this chapter is partially based on Salameh (1996) and Salameh (2004) with data updated from MWI (2016).

© Springer International Publishing AG, part of Springer Nature 2018
E. Salameh et al., *Water Resources of Jordan*, World Water Resources 1,
https://doi.org/10.1007/978-3-319-77748-1_7

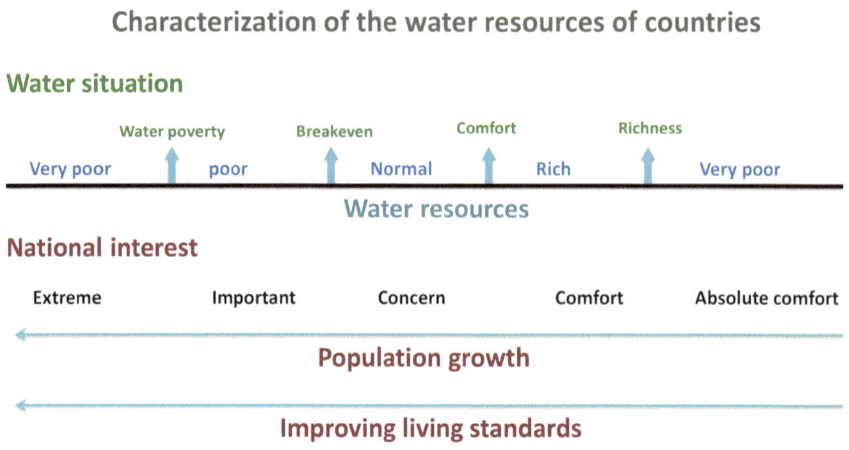

Characterization of the water resources of countries

Water situation

Water poverty	Breakeven	Comfort	Richness

Very poor	poor	Normal	Rich	Very poor

Water resources

National interest

Extreme	Important	Concern	Comfort	Absolute comfort

Population growth

Improving living standards

Fig. 7.1 Different national interest degrees in relation to the water situation expressing the poverty or richness of a country on water resources. The effects of population growth and living standards on the water situation are also illustrated

Water of the Jordan Valley in the political context
-Jordan-

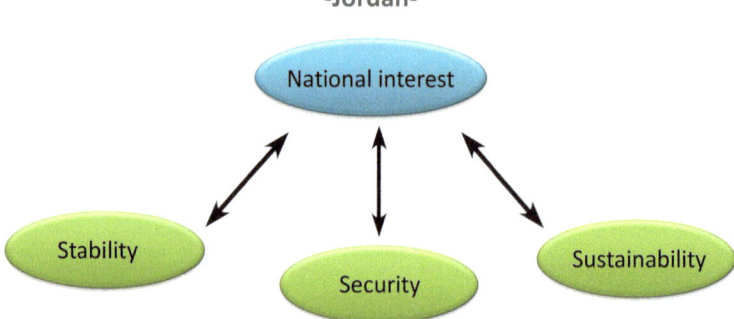

Fig. 7.2 Water of the Jordan Valley in political context

Fig. 7.3 National interest in the country's water resources as it is hierarchically reflected in the country's water policy, strategy, scenarios, action plans, programs and projects

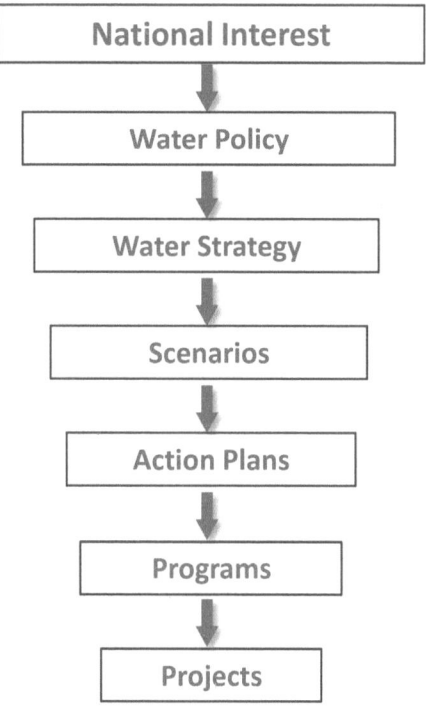

The three above mentioned national interests combine the social, economic, environmental and physical elements. All of them pour into one goal of development without jeopardizing the natural system; keeping it as a normally swinging pendulum and not allowing it to turn into an instable inversed one.

7.2 Water Policy Principles

Water policies consist of a set of broader outlines of principles to be adopted to serve the national interests. In Jordan these water policy principles can be summarized as follows:

1. The water resources of the country are a possession of the nation. The government and the parliament are responsible for their allocation to the different water use sectors and for their quantitative and qualitative conservation in accordance with the national interests.
2. Drinking water has a priority over all other water uses. Its quality must at all times and in all places be within the set standards.

3. All water resources of the country should be utilized to their utmost potentials under strict consideration of sustainability principles and preservation of biodiversity.
4. Differentiation between water use and water consumption should accompany any allocation of water to the different use sectors (water use means borrowing water and returning it to the environment, maybe with different characteristics; water consumption is the exhaustion of water with irrelevant return flows).
5. All water uses should be optimized (demand management), especially water use in irrigation which should be exposed to a strict comprehensive economic feasibility study of the water use itself and of the national economy as a whole.
6. Any exploitation of non-renewable water resources should be accompanied by national programs aiming at compensating next generations' rights in these resources. Such compensation can include investments to regenerate the exhausted resources at any time in the future.
7. Waste water should be considered as an integral part of the water resources. It should be treated to meet with the quality of the recipient water bodies or with the quality requirements of the next use.
8. All water should be priced according to its capital and operational cost in addition to the cost required to regenerate it if any part of it is non-renewable at the same rate of its exhaustion. This requires the optimization of the water supply to keep its cost optimal.

7.3   Water Planning

Background
To enable the development of water planning for an area profound, detailed knowledge and understanding of the water situation including resources, use, demands, quality, future needs, use sectors and areas, water rights, available water and waste water technologies, biodiversity, habits of population etc. are essential preconditions. Otherwise, it can not be called water planning.

Water planning here can be defined as the framework consisting of programs, projects, plans and measures to serve the national interests. The general policies of the water sector form the framework which sets the boundaries for the water planning to better serve the national interests.

Although different water plans can be advanced, their examination and comparison with and against each other facilitates choosing the most appropriate feasible water planning to fulfil the national interests by respecting the set policies, the time horizon, the available budgets and the available technologies.

In Jordan where the supply of naturally renewable water by precipitation is not sufficient to cover the requirements according to the country's Water Poverty Index, distinct differentiation between water use and water consumption has to be made. That is because water is not only renewable by the process of evaporation, condensation and precipitation but also by a variety of natural and artificial processes. Water is renewable by:

- Precipitation.
- Natural self-purification in rivers, lakes, during infiltration, ageing processes in lakes etc.
- Artificial treatment processes such as waste water treatment.

That part of precipitation which flows as surface runoff or which joins the groundwater is what conventionally has been considered as a renewable source of water. But, even in water rich regions of the earth this is not the only naturally renewable water, because, for instance, once used water (e.g. for washing) is returned to a recipient water body (river, lake) where it becomes exposed to self-purification processes, and it can be reused further downstream.

Seckler (1996) observes that consumed water is that water which is lost by evaporation or which is totally lost due to pollution (e.g. very high salinity).

Artificially, at reasonable cost, most once-used water can be recycled. This especially applies for municipal, food processing or cooling water.

The return flows of such uses can generally be treated and safely used e.g. in irrigation.

The differentiation between water use and water consumption leads to another term on the demand side, which is demand for consumptive uses.

In Jordan, consumptive uses in irrigated agriculture exhaust at present 85–90% of all the water made available for all the use sectors. 10–15% is consumed in households and industry. But, the actually pumped water for household and industrial uses amount to around 40% and 5% of the naturally occurring water.

7.3.1   Peculiarities of the Water Supply and Use System in Jordan

The water resources of each region of the world have certain characteristics in terms of how they are developed and used, their effects on population behavior etc. In Jordan these peculiarities can be summarized as follows:

1. Jordan has not only developed and used all its renewable water resources but development and use has broadened to exploit its non-renewable groundwater reserves which have been recharged and accumulating in the underground over hundreds to thousands of years. The non-renewable or fossil groundwater resources are now over-exploited by a rate of around 300 MCM/year. Since these

reserves are limited in their quality, extraction from then is quasi a mining process which will deplete them without rehabilitation in the next few decades.

2. The water supply of the major urbanized areas depends on developing remote water sources; Azraq, Disi, Wala, King Abdullah Canal and Zarqa Ma'in and Disi water for use in Amman and other central urban areas.

3. Procuring additional fresh water resources is highly problematic and it seems that demand management and reconsideration of the relevance of irrigated agriculture (especially that depending on fresh groundwater extraction in the highlands) has to be readdressed from both a socio-economic and environmental perspective.

4. The population of Jordan is growing at a rate of 2.4% per year. This growth rate is further straining the water situation. Water resources, supply networks, sewerage treatment plants etc. have to expand annually at this same rate just to ensure the same standard of services.

In addition, Jordan has historically had to supply water and provide sanitation for a huge number of refugees coming from Palestine, Lebanon, Iraq and Syria. These refugees have been and are still putting huge pressure on the country's water and sanitation sectors.

In addition, old systems have to be repaired, maintained and/or replaced. Next generations' rights in these sources must also be ensured. Hence, the earlier the stabilization in the population number, the easier water planning can be implemented.

5. The capital investment in dams, water supply networks, sanitation and sewerage collection and treatment systems is extensive and represents a burden on the financial capacities of most cities and towns and, in certain cases even those of the country. For all water-related infrastructures the necessary funds have to be reserved, earmarked and secured long-term before the actual measure is implemented.

6. Farmers used to get their irrigation water free or at a minimal price which only covered a small part of the running cost. The reasons for that are historic, dating from when intensive irrigated agriculture was mainly practiced in the Jordan Valley area in Jordan. That area became in the 1950s and 1960s the site of armed conflicts, which forced its population to move elsewhere. One of the incentives offered by the government to attract people back to the Jordan Valley area was to offer land and water for free or at very low prices.

7. The waste water produced in households is highly concentrated, because the per capita use of water is very low, about 80 liters per capita and day (L/c.d.) and at the same time the production of wastes per capita resembles that of other areas of the world. This situation creates major problems for waste water treatment facilities and personnel.

8. The water problems of the country are generally more severe than is commonly realized, even among decision-makers themselves. These problems are:

- Insufficient water resources
- The lack of water rights in the shared waters with other countries Water pollution
- Over-exploitation of groundwater resources
- Illegal use of drinking water in irrigation
- Illegal water extractions and
- Inability of the country to implement strategies due to repeated, unexpected refugee waves which enforce crises management in the water and sanitation sectors.

7.4   Recently Undertaken Strategies and Programs to Improve the Water Situation

The following Sects. 7.4.1, 7.4.2, 7.4.3, and 7.4.4 are based on Salameh and Udluft 2001.

7.4.1   Well Drilling Prohibition

In the early 1990s, driven by depleting aquifers and deteriorating water qualities, the Ministry of Water and Irrigation (MWI) started applying a new by-law which prohibited the drilling of new groundwater wells in areas where aquifers had suffered depletion and quality deterioration. The drilling of new wells was allowed only for the governmental drinking water supply, schools, hospitals, industrial plants and military establishments. The repair or substitution of damaged wells was allowed with the condition that the same specifications of the well to be substituted or repaired were maintained. This by-law was enforced for all companies dealing with drilling and supplies of well equipment, and stringent fines for violators were introduced. Due to the deep groundwater levels in Jordan, generally tens to hundreds of meters below ground surface, heavy machinery is needed for drilling. Such drilling equipment can be seen from far away. It is not like shallow drilling using light machinery or manual excavations which can be hidden in farms or behind constructions.

Impacts on Water Resources
Prohibiting the drilling of new wells meant stopping the acceleration of the depletion of resources and their quality deterioration. This does not mean that aquifers
will rehabilitate or recover, but that their lifetimes will be prolonged. The quantifi-
cation of the effects of well drilling prohibition on the groundwater resources is not
easy and requires tens of years to be recognized based on groundwater monitoring.
Estimates can be made by analyzing how much more groundwater would have been
extracted if well drilling had continued at the same rate as before the introduction of
the prohibition by-law. A few years prior to the introduction of the by-law, applica-
tions for drilling licenses had been around 5% per year of existing agricultural
wells. Therefore the groundwater extractions would have increased by 5% per year.
This means that the savings have been 5% per year over the last 20 years. Since the
over-exploited groundwater amounts are 300 MCM/year then at present we are saving (5%*20) 100% of the over-exploited amount each year, equivalent to 300 MCM/
year. The saved groundwater has remained in the aquifers.

7.4.2   Putting a Price for the Extracted Non-agricultural Water

In 1998 the government of Jordan implemented a new tool for the management of
groundwater resources which put a price on all extracted groundwater except water
used in irrigation. Groundwater extracted for industrial, commercial and municipal
uses, in schools, hospitals and military establishments must be paid for, although
well drilling, pumps, repair, maintenance and energy costs are paid by the users.
MWI provided all wells with water meters, read them on a regular basis and collected the bills.

Effects on the Groundwater Resources
The scarcity of water in Jordan also strongly affected industries, which during the
dry season were not able to obtain enough water even from their own private wells
due to depletion and in some areas the water quality deteriorated beyond limits at
which it could be used. This together with the by-laws to meter and charge for
extracted water forced some industries to introduce water saving devices whenever
and wherever possible. The recycling of water within industrial plants was also
introduced. The savings in the industrial sector as a result of metering and paying
for extracted water, installing water-saving devices and recycling once-used water
are estimated to be around 10% or about 5.5 MCM/year of the water amounts used
in industry. At present it does not seem possible that industry in Jordan can save
more water by additional savings and recycling.

7.4.3 Metering the Groundwater Amounts Abstracted for Agricultural Uses

In 1999, the Ministry of Water and Irrigation introduced water meters on all agricultural wells. This action plan had two aims, namely:

1. To measuring the amounts abstracted from each aquifer.
2. To remind farmers to abstract only the amounts of water stated in their well drilling license.

Effects on the Groundwater Resources
Due to the complexity of factors effecting the groundwater levels and amounts such as amounts of precipitation, natural recharge extraction etc. the impacts of saving cannot be easily measured, but due to the decreases of the extracted groundwater amounts, declines in the depletion and salinization rates of aquifers must have occurred and the life of aquifers prolonged.

7.4.4 Pricing Water Extracted for Irrigational Uses

In early 2000 the Ministry of Water and Irrigation issued new regulations according to which extracted groundwater for agricultural uses would be charged for water amounts exceeding the requirements of family businesses or the subsistence economy. A block-type of tariffs was introduced according to which water prices increased with increasing amounts of extracted water from a well. The metering and pricing regulations have since then limited the amounts of extracted groundwater and based on that, have saved appreciable amounts of groundwater, which has alleviated the over-exploitation and deterioration of aquifers.

7.4.5 Dams, Treated and Untreated Water Storage

Dams in Jordan such as King Talal Dam (KTD), Shueib, Kafrain were constructed to collect flood and base flow water for use in the dry season and to increase the availability of water in the country. Within the catchment area of these dams waste water treatment plants were also established to treat household and industrial water, which is a sound policy measure to protect the environment in a country suffering from water shortages and where waste water can be considered as a relevant water source for reuse after treatment. But, the effluents of the waste water treatment plants of es–Samra (Zarqa River), Salt Wadi Shueib, Fuheis Wadi Shueib and Wadi Sir (Wadi Kafrain) were then allowed to flow into the reservoirs of dams constructed along the discharge wadis. This action, although increasing the stored amount of water in the respective dams, has caused deterioration in the original water quality of the dams, which formerly consisted of good quality base and flood flow water. The most important detriment to the water quality was the initiation of eutrophication processes rendering the dams' water unsuitable even for unrestricted irrigation let alone drinking purposes. Treatment of waste water was not always sufficient and the waste water accumulated all the junk that had been dumped in the valleys during the dry season.

In the case of KTD, lands irrigated by its strongly eutrophic water in 1991 ended up with damage totalling 60,000 dunums (6000 ha) of crops. The dam's water, at that time deprived of dissolved oxygen, containing H2S and a high concentration of toxic trace elements, released eutrophication processes from the bottom of the lake. The eutrophication processes resulted in microbial blooms which turn the water into a green substance. The highly concentrated algal cells sank to the lake bottom and their decomposition by bacteria used up all the oxygen so that life in the bottom substance (mud) was limited to only *Archea* (anaerobic) bacteria which produce methane and H_2S, both poisonous to aerobic life.

The water was conveyed through closed pipes and applied to plants via drip irrigation with no opportunity for aeration. Therefore, the plants suffocated.

The policy of storing treated and in some cases untreated waste water in dams designed to store fresh water has caused the serious deterioration of the quality of water in these dams.

It is now high time to reconsider that policy so that treated waste water effluents are stored away from dams whether planned or built to store good quality flood and base flow waters. This means that treated waste water should be stored separately from water coming from other sources.

7.4.6   Deep Wells at the Escarpment Foothills of the Jordan Valley

In the 1980s many deep wells were drilled at the foothills of the escarpment overlooking the Jordan Valley. These wells encountered brackish artesian water from the Jurassic – Triassic aquifer, and in some places mixed with Kurnub (Lower Cretaceous) aquifer water. In order to evaluate this area it has to be taken into consideration that it has been ruptured by many faults on the margins of the Jordan Dead Sea and Wadi Araba Graben, many of which are still active.

The problem started when the drilling companies were not able to stop the artesian discharge of these brackish water wells. Neither has the water been used for any purpose, nor has the discharge of the wells been stopped. Moreover, the discharged brackish water flows along perennial fresh water courses causing the original water quality of these water courses to deteriorate and the environmental conditions to degrade. Such wells were drilled in the wadis Hisban, Kafrain, Rama, and at other sites.

In addition, these artesian wells also discharged water from aquifers extending eastward under Jordan. The water contained in them forms the backbone or the base support of all the overlying fresh water aquifers, supplying fresh water for the different use purposes of drinking, industrial, irrigation and tourism.

When water is released from the deep aquifers through the deep wells drilled at the foothills of the escarpment, ground water from the overlying fresh water aquifers will flow downwards to substitute the deep aquifer water released from the artesian wells. The downward moving fresh water hence becomes brackish due to the saline nature of the deep aquifers. The results are then decreases in the amounts of stored fresh groundwater in the shallow aquifer, a drop in the level of the fresh shallow groundwater and loss of fresh water. This is in addition to the above mentioned salinization of the fresh water of wadis, along which the brackish water of the artesian well flows, which makes these wadi waters useless and detrimental to the eco-system in the downstream areas along the wadis. Seen from the perspective of a scientist in hydrogeology, this represents a criminal act against the resources of fresh groundwater of the country, and hence against its economy, water supply, and the healthy survival of its population.

These wells should by any means be refilled to the depth of the producing geologic units in order to prevent further fresh water losses from the overlying aquifers and to stop further salinization of fresh water along the courses of the respective wadis. But even if some of the wells stopped flowing at the ground surface, they will continue emptying the deep aquifers by flowing upward in the drilled wells into the shallower aquifers through the corroded casings due to dissolutions caused by the acidic brackish water containing H^2S and other aggressive gases. Thus it will be extremely difficult to rectify former mistakes, and more sound initial studies of the geology of a proposed drilling site should be carried out in the future before drilling takes place.

7.4.7 Irrigation in the Highlands Using Pumped Fresh Groundwater

The drilling of wells along the highlands and in the eastern parts of the country started as a government project, for the purpose of supplying drinking water and allowing Bedouins to settle in these regions. Afterwards well drilling licenses were granted so that personal businesses, i.e. family projects, could be established. After that companies began to obtain single drilling licenses for wells to irrigate large areas of land of about 40,000 dunums (4000 ha) per farm.

The speed with which well drilling licensing and activities were performed was amazing. Around 4000 wells were drilled in less than two decades (1970s and 1980s), many of them without scientific analyses being carried out beforehand. A large number of illegal (unlicensed) wells were also drilled in all parts of the country.

Groundwater has not only been extracted from renewable aquifers, but also from non-renewable aquifers by drilling deep wells of more than 300 m.

As a result of excessive groundwater extraction, particularly for irrigation purposes, groundwater levels started to gradually drop and water quality to deteriorate. The extracted groundwater amounts are presently estimated at 550 MCM/year, whereas the safe yield of groundwater extraction in Jordan should be around 275 MCM/year according to the results of the National Water Master Plan studies 1977, updated in 1998 and afterwards.

The impacts of groundwater over-exploitation have become detrimental to the country:

- Groundwater levels have been dropping dramatically by 20–150 cm/year.
- Groundwater quality has deteriorated due to salinization and the mobilization of salt and brackish water bodies.
- In some areas, such as Jafr, Dhuleil, Za'atari, Agib and parts of Azraq basin, the groundwater has degraded to a point beyond repair.
- Due to the over-exploitation and emptying of aquifers, the government in no longer able to supply drinking water in suitable quantities and qualities and has to resort to big expensive projects that are beyond the financial means of the country.

The catastrophe that has been inflicted on the groundwater resources of Jordan is the result of:

- Inadequate studies to evaluate the water resources of the different basins and aquifers and to allow extracting amounts fulfilling the safe yield principles.
- Inadequate law enforcement with regard to illegal well drilling.
- National interests were compensated for personal interests in the context of agricultural business based on non-renewable and fossil groundwater resources. These agribusiness projects import fuel, machinery, seed, biocides, workers etc. and the only input in these irrigation projects from Jordan is the precious, non-renewable groundwater, knowing that Jordan is among the poorest countries in the world in respect to its water resources.

Agriculture irrigated by fresh groundwater in the highlands and eastern and southern deserts should be gradually curtailed, because it is damaging to the groundwater resources of the country. Although it may be economically beneficial to farmers and agribusiness, it is a burden on the national economy, because the extracted groundwater cannot be replenished and its actual value is equivalent to the value of desalinated water plus the cost of pumping the desalinated water to the present extraction areas of the groundwater.

Deir Alla – Zai – Amman

Parallel to the Jordan River, along the eastern side of the Jordan Valley the King Abdullah Canal (KAC, formerly East Ghor Canal) was constructed in the 1960s to convey water from the Yarmouk River and from the Jordan Valley side wadis for use via natural flow by gravity e.g. no pumping in irrigating lands at the eastern side of the Jordan Valley.

At the end of the 1970s as a result of water shortages in the capital city Amman a decision was taken to pump water from the KAC at Deir Alla to Amman for use as drinking water. Other alternatives were suggested to supply Amman with water, such as what came to be known as the Mujib–Zara–Amman project.

The problems associated with the Deir Alla–Zai project were analyzed by experts and the government was advised not to proceed due to:

1. Doubts about the treatability of the KAC water suffering of strong eutrophication processes.
2. Development of carcinogenic substances during treatment and application of chlorine disinfection.
3. Inadequacy of the planned treatment and procedures (at that time).
4. The irrigation sector in the Jordan Valley is badly in need of the KAC water.
5. Existence of other relevant and more secure alternatives.
6. If the decision is to continue taking water via the Deir Alla–Zai project, then the water should be taken directly from the Yarmouk River via a piping system and not from KAC at Deir Alla when the Yarmouk water has flows for 65 km in agricultural land with all possible pollution sources along an open canal.

Implementation of the project had the following consequences:

1. In 1987 a major catastrophe hit Amman and its surroundings when its drinking water became polluted and thousands of people suffered as a result of the bad odor and taste of the water and the parasites contained in it. The project was suspended for months to improve treatments and to add activated carbon filters. As a result of this catastrophe the presidents of both WAJ and JVA were released from their positions.
2. In 1998 a similar catastrophe occurred and the population of Amman and its surroundings again suffered from illness and from the bad taste and odor of the polluted drinking water. The project was stopped again, and as a consequence the government was resolved.
3. From 1987 until the present, the project has encountered pollution problems at least once a year, and the water supply of Amman has been interrupted, putting engineers, water specialists, laboratory staff and policy–makers under severe stress, as well as adversely affecting all the officials concerned and the population supplied with that water.
4. Ever since the Deir Alla–Zai project came into operation, farmers in the Jordan Valley have been deprived of the quantities of irrigation water necessary for their

crops, although the KAC had originally been established for irrigation purposes.

The drinking water supply from the Jordan Valley water, especially due to the decrease in the Yarmouk River flow as a result of Syrian extraction during the last decade, has been strong competition to irrigation water in the Jordan Valley. Local farmers have gradually started to criticize the taking of that water to supply drinking water to Amman.

It has become quite clear that the Deir Alla–Zai drinking water supply has negative socio-economic impacts on farmers in the Jordan Valley. It also has negative health impacts on the population drinking that water. And due to its high trihalomethane concentrations it is suspected of being carcinogenic.

Treatment at Zai produces refuse water which discharges to Wadi Haramiya which then discharges into KTD and causes further deterioration in the dam's water quality.

Politically, pumping water from the Jordan River system, of which the Yarmouk River (Deir Alla water), is a part, seems to justify the Israeli pumping from Lake Tiberias into the National Water Carrier of Israel. This is in the knowledge that pumping from the Jordan River system is not within the Unified Johnston Plan of 1955, agreed upon by all the Jordan River system riparian countries and by the Arab Liege, the agreement sees for water use only by gravity flow (no pumping allowed).

7.5 Water Resources in Regional Context

It was clearly demonstrated in the last section that Jordan is severely strained in terms of its water resources in relation to population and development dynamics. The main water resources of Jordan are shared with neighboring countries.

The shares and rights of Jordan and neighboring countries are spelled out in the Johnston agreement, which was not formalized but was informally accepted with regard to the bases of water rights for Jordan, Syria, Lebanon and Israel. As Table 7.1 indicates the water share for Jordan was 477 MCM/year in total (100 from Jordan River, and 377 from Yarmouk River). However, the implementation of this agreement was never fully realised and Jordan has not received its share due to the Arab-Israeli conflict and Syrian intransigence.

As shown in Table 7.2, the actual amount of water that Jordan recieved was only 29% of its total share. This indicates that it is extremely difficult to solve the water scarcity problem in Jordan without regional consideration or solutions.

This idea is also supported by the political developments in the region since the 1950s and also the development and utilization of new technologies and resources both in Israel and Syria.

Table 7.1 Water shares of riparian countries of the Yarmouk and Jordan Rivers according to the Unified Johnston Plan (MCM/year)

River	Jordan	Syria	Lebanon	Israel	River total
Jordan	100	42	35	375	552
Yarmouk	377	90	00	25	492
Country total	477	132	35	400	1044

Table 7.2 Water use of the riparian countries of the Yarmouk and Jordan Rivers compared to shares according to the Unified Johnston Plan (MCM/year)

River	Jordan	Syria	Lebanon	Israel
Jordan	**100**	**42**	**35**	**375**
	30	00	25	650
	(−70)	(−42)	(−10)	(+275)
Yarmouk	**377**	**90**	**00**	**25**
	30	330	–	25
	(−347)	(+240)	00	(00)
Country total	**477**	**132**	**35**	**400**
	60	330	25	675
	(−411)	(+198)	(−10)	(+275)

Israel's reliance on water resources from the Jordan and Yarmouk Rivers has changed because of the increasing expansion of the water desalination project in the Mediterranean Sea. Also, Jordan, Israel and the Palestine National Authority (PNA) are cooperating on the implementation of the Red-Dead Canal project where certain agreements have been reached for sharing the desalination water, the outcome of which will set the stage for future discussion on other issues.

In regard to Jordan's water share from the Yarmouk River, Jordan was only receiving 14% of its allocated share in the Johnston Unified Plan. Syria has been undergoing changes since 2011 and the civil war that resulted in the near collapse of central state authority over vast territories including the Yarmouk River area. This development, and based on the final solution of the Syrian crises, could present new opportunities for Jordan to realize it's rights regarding the Yarmouk River.

7.6   Resource Shortage in Jordan

In sum, regional cooperation is a key to the solution of water shortage in Jordan and other countries in the region. The declining salience of the Jordan River water resources allocated to Israel, combined with desalination water abundance and the decline of central state authority in Syria might present a fresh perspective for revising the water resources management system among these countries.

7.7   Water Environmental Services

Natural surface water bodies, such as creaks, rivers, ponds etc. provide special habitats for a variety of organisms and species, and are sanctuaries for migrating birds.

Springs in Jordan used to flow along wadis, especially those flowing in a westerly direction towards the Jordan Rift Valley and to the east to collect in the Azraq Oasis or playas, which dry out in the summer season.

Projects aiming to supply drinking water by capturing spring water diverted all the springs' water along almost all the wadis in the country and left no water for the environmental services of the water along the wadis.

This policy has led to the drying out of wadis during the dry season and during extended dry periods during the wet season, to the death and extinction of many species of their flora and fauna and to the almost complete destruction of their natural habitat and local ecology.

Under any circumstances such a policy can be described as criminal with regard to the country's ecological system. It is now high time to take natural habitats and ecosystems into consideration, not only to protect the few springs which are still providing wadis with water but also to rehabilitate the wadis by leaving some water to flow again for the purpose of water environmental services.

The few water courses whose original environments are still found intact are in desperate need of protection. Such water courses are: Hisban, Wadi Shita, Wadi Atoun, Wadi Mujib downstream of the dam and to a certain extent Wadi Ziqlab.

Water courses strongly affected by the total water diversion and by pollution are Shueib, Zarqa and its wide mesh of tributaries.

Already at the beginning of the 1980s the different water courses of a large number of relevant wadis were studied on their environmental health and their environmental services, and recommendations to protect them were advanced. Nevertheless, the situation since then has not improved but on the contrary, regrettably worsened.

Karama Dam
The following section about Karama Dam is based on Salameh 2004.

The rocks building the catchment area consist of the recent alluvial deposits of the Jordan Valley area, which at shallow depths are underlain by the salty Lisan Formation deposited in the Lisan Lake, the ancestor of the Dead Sea.

Geologic studies on the area of Karama Dam show that the dam site is located in a major weakness zone of the earth's crust, where the major transform fault of the Dead Sea structure runs (Bender 1968; Brink et al. 1999). The site is also a major focal point of earthquakes, which date back to hundreds of years (Abu Karaki 2000).

During the investigation of the dam site, some scientists and specialists advised the government not to proceed with the construction of the dam because of the:

– Geological instability of the site
– Existence of major faults passing through the dam site
– Seismic risk

- Salinity of the rocks underlying the dam and reservoir sites
- Presence of salt water springs at the bottom of the dam
- Sources of water to fill the dam are not clear

Presence of other suggested sites for dam projects with higher priorities compared to Karama dam (Salameh and Teimeh 1992). But in spite of all warnings the Jordan Valley Authority (JVA) of MWI went ahead and started construction of the dam.

The design storage capacity of the dam reservoir is 55 MCM and the catchment area feeding it is 61.2 km². Precipitation over the catchment averages 150 mm/year and the annual potential evaporation is 2450 mm/year.

The catchment area of the dam is almost flat and covered by the recent deposits of the Jordan Valley area with their high porosity and permeability, which means that surface runoffs resulting from rainfall are very limited if any. At the dam site they are estimated at an average of 1 MCM/year. Hence it was planned to fill the dam from KAC, which is mainly fed by water coming from the Yarmouk River.

Since its construction 21 years ago the dam has failed to collect water from within its own catchment. The fresh water diverted into the dam reservoir from King Abdullah Canal to fill it and to demonstrate its success turned out to be highly saline (15–20 g/L). Reservoir bottom and abutment collapses have been common. Equipment and facilities are corroding and deteriorating and the government has been covering depreciation. Capital and running costs of the project since its construction have also been born by the government with absolutely no returns and no use of the Yarmouk River into the Dead Sea. Not a single drop of the water collected in that dam has been used for any use purpose.

It was a hard lesson for the government of Jordan, spending around $75 million to construct the dam and losing around 700 million cubic meters of fresh water which has been used since its construction to wash the dam's reservoir of salts with no use of the collected water in the dam. Here it must be stressed that the companies that were consulted and recommended the construction of the dam must also be made responsible and accountable for its failure. It is concluded here that accountability and liability insurances must be integral parts of the contracts.

7.8 Conclusion

This conclusion is partly based on Salameh and Udluft (2001).

After developing all potential water resources of the country, whether surface or ground, and treating and reusing waste water Jordan is still in desperate need of additional water to cover the demands of the population, which is increasing as a result of natural growth and refugee waves coming from surrounding countries. Within the country only expensive water-saving projects in agriculture can postpone Jordan's crisis a little longer. Plans to limit agricultural uses to their present level, restricting household water use to the least amount necessary for hygiene, covering

cover only the demand for satisfy the needs of natural population growth, and giving industries the minimal amounts of water to cover their needs will not stop the water problems of Jordan in the short-term let alone in the few years' time.

The policy of the government of Jordan according to which the capital and running cost of projects to allocate water for irrigation such as dam, canal and water harvesting projects and subsidizing irrigation water, are encouraging water wastage. However, farmers would irrigate their fields more efficiently and introduce water saving techniques and devices if the irrigation water was priced at its capital and running cost. Pricing water low has led to wastage and to the inability to satisfy demand. Farming companies using non-renewable and fossil water in irrigation in Azraq, Dhuleil, Disi and other areas pay only the pumping costs of the groundwater, but not the pumped water itself, which will, in the mid-term, certainly lead to the depletion and salinization of the groundwater resources and hence to the loss of the nation's future water and food security. Paying the cost of water allocation and a basic price for water as a natural resource even for water used in irrigation will certainly lead to the saving and conservation of part of the extracted water and to reconsideration of the economics of many agricultural projects.

In the coming decades only high-cost projects such as desalination of sea water will make additional water available for the different use sectors of the country. Therefore, policy-makers have recently started to look at desalination as the major water development project to solve the country's water shortages. Lowering the demand for water and increasing the efficiency of water use instead of increasing the supply will not add the envisaged amounts to the water supply of the country.

Competition for water resources started in Jordan as a result of limited resources and increasing demand due to population growth, higher standards of living and influxes of refugees.

The water sector in the country now requires new thinking and new management procedures. Social and political issues which determined water use allocations in the past are now obsolete. Water should now be considered as an economic good. Accordingly, allocating certain water sources for agricultural activities may have to be abandoned. But that may cause some difficult socio-economic and political problems such as more expenditure to import food, unemployment and poverty leading eventually to social unrest.

The increasing demand for water and the limited resources have made it necessary to think outside the box and have made the management and development of the water sector through piecemeal projects irrelevant and insufficient. Therefore, the Royal Committee on Water together with the Ministry of Water and Irrigation were commissioned with the development of a dynamic water strategy for the country which has been accomplished with adequate dynamic instruments in it to enable comprehensive planning and project implementation. This serves now as a reference for donors willing to assist the water sector.

Technological progress during the last four decades means that we can now develop a more efficient water supply system. This process started some 15 years ago and has intensified in the time since. Improving water economics is certainly not an easy process in a traditional society that thinks that water is given by God and that it should be available that thinks that water is given by God and that it should be available free of charge. Yet improving water economics has allowed more funds to be allocated to increase the efficiency of water supply systems; it has also allowed better detection of misuse and the control of water systems curtailing water theft and alleviating physical water losses, which has resulted in water savings and thus more water becoming available for use.

The prevailing shortage in water resources over three-four decades and the expected sharpening of demand while water resources remain limited has initiated the introduction of more efficient conservation systems rather than searching for new resources. Over the last 15 years Jordan has introduced and developed the necessary technologies for a better water supply system and better waste water treatments and reuse schemes. These technologies along with new managerial measures have during the last 15 years alleviated the shortages in the water supply, but the increase in population, especially the influx of refugees has made Jordan's water challenge more difficult. The traditional policy of developing new resources to satisfy needs is, in the case of Jordan, totally exhausted. Now is the time to look for non-conventional water resources to cover the increasing demand for household uses, industry and tourism. Agriculture should not receive additional fresh surface or groundwater. Its additional allocations should only come from treated waste water which increases with increasing household water uses, especially because treated waste water has become an integral part of the water resources in the country. Waste water therefore has to be adequately treated to make its application in irrigation safe.

Desalination, obtaining Jordan's fair share in the water resources shared with other countries and obtaining additional regional water from Jordan's northern and north-eastern neighbors may offer a better water future for the country.

Recently, a new opportunity has opened for the countries of Jordan, Israel and Palestine represented by Israel, namely the desalination of increasing amounts of water from the Mediterranean Sea for the use of inhabitants of this part of the world, making the water of Lake Tiberias available for other uses. It is an open opportunity for Jordan and Palestine to reach an agreement with Israel to utilize that water for their needs. This opportunity is readily available, cheap to develop and can provide an appreciable amount of water for use by the two countries.

In addition, Syria had been, until the onset of the internal conflict, using some 260 MCM/year of Jordan's share in the Yarmouk River water. But, deteriorating conditions in Syria made the use of the water stored in reservoirs in Syria a difficult enterprise. In addition, due to the internal conflict the maintenance and repair of water use and storage facilities has become very difficult. These two facts have put Jordan in a better position to re-negotiate obtaining more of its water share from the Yarmouk River (Table 7.3).

Table 7.3 Additional inland water resources and the different aspects of making them available

Project	Required structures	Cost (estimated) million	Time required to build	Complications	Treatment	Exploitable amounts	Environmental impacts	Socioeconomic impacts
Flood water harvesting	Mainly weirs	Very low	Is going on; Weirs require short time for construction	Tribal fronts and water rights	None for irrigation and cattle watering, filtration and disinfection for drinking	Around 20 MCM/year	Positive if some water is left to flow along wadis for environmental services	Very positive for Bedouins and animal husbandry
Saving & reallocations	Pipeline according to site, compensation of present users in irrigation	Depend on site, relatively cheap	A few weeks to months	Water rights of present uses, willingness of farmers to lease or sell or exchange for treated effluents	None	200–250	No major impacts especially if the treated wastewater is reused.	Positive on drinking water supply, but has to be studied in each case, very minor in most cases. If severe in places it should not be implemented.
Yarmouk	Minimal	Negligible	Negligible	Political: the Syrians must be convinced to release Jordan's share of water from their dams	Filtration + disinfection	80	Positive	Positive
Brackish water	Pumping stations, pipes, eventual desalting or mixing.	10–20	1–3	Environment; Discharge of desalination brine	Mixing or event. Desalting	Ca. 30	Positive, if the brine is discharged into the Dead Sea	Positive: additional return flows for irrigation

Cost in $US, time for implementation in years, exploitable amount in MCM/year

References

Abu Karaki LO (2000) Skeletal biology of the people of Wadi Faynan: a bio archaeological study. Unpublished MA thesis, Faculty of Archaeology and Anthropology, Yarmouk University

Bender F (1968) Geologie Von Jordanian. Beitrage zur Regionalen Geologie der Erde, vol 7. Gebruder Borntraeger, Berlin

Brink US, Uri S et al (1999) Anatomy of the dead sea transform: does it reflect continuous changes in plate motion? Geology 27(10):887–890

Brundtland GH (1985) World commission on environment and development. Environ Policy Law 14(1):26–30

MWI (Ministry of Water and Irrigation) Jordan. Open files (2016)

Salameh E (1996) Water quality degradation in Jordan. Friedrich Ebert Stiftung, Amman and Royal Society for the Conservation of Nature, Amman, 179 p

Salameh E (2004) The tragedy of the Karama Dam project. Acta Hydrochim Hydrobiol 32:3

Salameh E, Teimeh A (1992) Internal report in the University of Jordan

Salameh E, Udluft P (2001) "Towards a water strategy for Jordan" Hydrogeologie und Uwelt, Wurzburg University, Germany

Seckler DW (1996) The new era of water resources management: from "dry" to "wet" water savings, vol 1. Iwmi, Colombo

Chapter 8
Conclusions and Recommendations

8.1 Resources

Water availability in the Middle East and North Africa has, throughout human history over the last few millennia, largely shaped human life and lifestyles, their socio-economies and even their conflicts (Salameh and Udluft 2001).

Rain-fed agriculture was practiced where the amount of precipitation allowed plant growth and irrigated agriculture developed along perennial water courses such as the Jordan, Yarmouk and Zarqa Rivers as well as around the few water pools in the plateau area in the eastern part of the country.

Jordan's development between the 1950s and 1990s concentrated on agriculture, especially irrigated agriculture, which incorporated the developing of water resources for use in irrigation. The aim was the creation of job opportunities for Jordanians and refugees. The capabilities of Jordanians and refugees and the stage of the society's development at that time allowed only for the development of agriculture to avert the potential catastrophes of poverty and hunger. It also facilitated the settlement of refugees and local nomads and fostered the social, economic and political development of the country. Further development of the agricultural sector occurred with each new influx of refugees from Palestine in 1948, 1967 and partly in 1991 after the first Gulf War. In the face of the sharp increase in the population due to natural growth and refugee waves from Iraq and Syria after 2003 and 2011 in addition to those from Palestine, as well as the establishment of many industries such as potash, phosphate and fertilizers, the available water resources were no longer sufficient to meet demand. Between the 1950s and 1960s and the present day the water use sectors in Jordan have widened from low supplies per capita for drinking (20–40 l/c.d.) and for irrigation to higher uses for drinking (80 l/c.d.) and alsoto satisfy the demand for the developing industries, tourism and recreational water uses. The gap between available resources and demand has also strongly widened, forcing the country to resort to water desalination (Salameh and Udluft 2001).

Despite the implementation of intensive water projects and reservation measures, water shortage has become the major obstacle to Jordan's development. This is putting water experts and politicians under severe stress with regard to the future of the

---

The following conclusion on resources, Sect. 8.1 is based on Salameh and Udluft (2001).

© Springer International Publishing AG, part of Springer Nature 2018   143
E. Salameh et al., *Water Resources of Jordan*, World Water Resources 1,
https://doi.org/10.1007/978-3-319-77748-1_8

country's economic growth, especially since the water problems of the country are numerous, including: increasing demand for the local population and refugees, limited and, depleting resources, over-exploitation, the exhaustion of non-renewable resources and pollution.

The prevailing climate in Jordan is semi-arid. Only the highlands in the west and north-west can be characterized as Mediterranean. Jordan receives an average yearly amount of precipitation ranging from 30 mm in the south-east and east to about 600 mm in the north-west.

The evaporation force of the climate in Jordan is very high: in the cooler north-western areas, it is about 1800 mm per year and in the south-east it goes up to 4200 mm. This is respectively, three and 140 times the amount of average annual precipitation.

Perennial water in Jordan is found mainly in the rivers and wadis of Yarmouk, Zarqa, Mujib, Zarqa-Ma'in and Hasa. These discharge water during all seasons into the Jordan River, the Dead Sea and Wadi Araba, with its ultimate destination as the Dead Sea. In addition to rivers and wadis, the Azraq Oasis, situated 100 km to the east of Amman, used to hold water in all seasons. These sources, excluding the jointly-owned Yarmouk River, discharge approximately 160 million cubic meters annually, less than the average discharge of the Nile in one day and less than that of the Euphrates in two.

The groundwater resources of the country are of two origins: (1) recent and renewable and (2) fossil, which receives no or only a very small amount of recharge. The latter is non-renewable in technical terms and its exploitation is equivalent to a mining process. The fossil ground-water resources are mainly found in the southern and eastern parts of the country. They infiltrated into the aquifers tens of thousands of years ago, when the prevailing climate was more humid. Such water can be considered a reserve for dry years.

The renewable groundwater resources of Jordan excluding the Yarmouk basin, amount to about 340 MCM/year. Up until the 1990s they used, to suffice for the greater part of domestic and industrial needs and cover some agricultural uses. After that, non-renewable and fossil groundwater had to be tapped to cover increasing demands.

8.2 Projects

As mentioned above, water-resources development is of great concern and a priority for the country. Dams were constructed, irrigation canals were built, and domestic water supplies were extended to serve 98% of the inhabitants including those in the remote and sparsely populated areas of the country. Even in areas where the source of water was tens of kilometers away from the settlement, water was delivered to the inhabitants through pipe connections. More than 30 cities and towns (63% of

Table 8.1 Water supply of Amman from far resources, their distances, pumping quantities and pumping heads

Water source	Distance to Amman km	Pumping head m	Water quantity MCM/year
Disi	320	100	100
Azraq	110	650	20
Wala	55	250	22
Mujib Zarqa Ma'in	90	1500	45
Deir Alla (KAC)	55	1400	60

Jordan's population) are now served by sanitary sewerage systems and waste water treatment facilities and another 30% by improved waste water disposal facilities.

In the Jordan Valley area, KAC (formerly the East Ghor Canal) was constructed along the eastern bank of the Jordan River. It extends some 110 km and irrigates 170,000 dunums. Other irrigation projects were implemented in the southern area of the Dead Sea, putting around 46,000 dunums to use. In addition, the lands of the Jordan Valley lying above the reaches of the canal were irrigated using the waters of the side wadis and some groundwater, bringing the total irrigated land in the Jordan Valley to around 280,000 dunums.

Concerning domestic water supplies, expensive projects proved to be necessary in order to serve the population centers, which are generally located away from potential water resources. For example, the capital city of Amman gets its domestic water from various sources as detailed in Table 8.1.

The distances between sources and use centers and pumping heads are, as can be seen from Table 8.1 for a non-oil-producing country, a very expensive affair.

8.3 Water Use and Resources Development

The indigenous population of Jordan had been growing at the high rate of 3.6 % per year during the 1960s–1990s which has decreased since then to 2.4%. Accordingly, the population of Jordan is expected to grow to 7.3 million by the year 2020 and to 8.8 million by 2030, i.e., to double in 20 years. If living standards and population structure remain at their present state, domestic water use is also expected to double in the same time period. Any rise in living standards or social-structure order will result in higher water demands, which will exceed double the present consumption.

The present per capita daily water use is 80 liters. Of the present total amount of water pumped to consumers; 220 MCM/year, one quarter is lost through corroded leaky pipes, another quarter is used illegally (not being paid for), and a fraction is used by small-scale industries.

The planned industries are also expected to consume more water. The demand is calculated to rise from around 57 MCM/year at present to 75 MCM/year by the year 2020 and to 110 MCM/year in 2030.

Refugees

It is estimated that between 550 and 600 MCM/year of water were used for irrigation over the last few years, distributed between surface and groundwater resources and treated waste water.

Added to the domestic and industrial consumption, the total water use amounts to 1057 MCM/year. The total extractable and renewable water resources of the country are around 896 MCM/year. It is worth mentioning at this point that almost all the groundwater resources are at present over-exploited, such as Dhuleil, Azraq, Disi, and Wadi Arab, Mujib, Zarqa Ma'in and Yarmouk. And in general, the water resources yet to be developed are very meagre, suffer from salinity or are partly shared with other countries.

Even if the amount of water used for irrigation is limited to its present level, and if water projects and extractions are redistributed in line with the safe-yield concept, Jordan is now using all of its available and renewable resources. Not only were renewable water resources used, but extractions were expanded to include the fossil and non-renewable water resources which have been stored underground for thousands of years. Some of these resources have been exhausted because their replenishment rates cannot cover the extraction rate. This is the case in Dhuleil, Jafr, Mafraq, Za'atari, and Agib. The resources of Azraq, Mujib Sultani, Qatraneh and some Disi areas are now also threatened by depletion and salinization.

Failure to plan water projects carefully has resulted in exhaustion or damage to some sources.

8.4   Pollution and Over-Exploitation

During the last four decades small and medium-sized industries have been established in Jordan, concentrated mainly in the Amman-Zarqa area. Effluents from some of these industries are only partly treated and are directly discharged either into the nearby wadis or into the sewerage system, causing the deterioration of surface and groundwater quality or retarding treatment in waste water treatment plants. This type of pollution is limited in its distribution and extent, and major regulations and practical steps have been taken to alleviate its effects.

The major pollution problems during the 1980s up to early 2000 were the result of inadequate treatment of domestic waste water in inefficient waste water treatment plants, the choice of inferior waste water treatment plants and inappropriate reuse schemes. After constructing new, more appropriate municipal waste water treatment plants the environmental situation in Jordan has greatly improved.

The still existing cesspools and the leaky sewers cause pollution of surface and groundwater resources. Cross-connections with the leaky water supply net is, in some places, also leading to contamination of municipal water supplies mainly as a result of interrupted pumping and the resulting negative pressure in the supply network.

Jordan's scarce water resources, lack of perennial flows, hot climate and relatively low per-capita use of water result in a dense waste water with highly concentrated pollution parameters, which renders the current choice of treatment plants
and technologies inadequate. The insufficiently treated effluents are not diluted due
to the scarcity of perennial water such as rivers. The toxicity of effluents and the hot
climate accelerate eutrophication processes in surface-water bodies, rendering the
main reservoirs highly eutrophic (ageing lakes). The effects of treatment-plant
effluents are also damaging to the groundwater resources, especially the effluents of
less effective waste water treatment plants. In the past waste-stabilization ponds
have proved to be unsuitable for countries with poor water resources, where sewerage is very concentrated, evaporation rates are high, and no dilution takes place.
Added to this problem is the extremely unwise choice of treatment sites such as
Baqa'a, Madaba, north Irbid and Salt. Also, the wrong choice of solid waste disposal sites and methods has led to the deterioration of both surface and groundwater
in their areas.

Over-exploitation of aquifers on account of the country's reserves of non-
renewable and fossil water is gradually leading to aquifer depletion and
exhaustion.

In certain areas, Dhuleil Jafr, Azraq, Za'atari, Agib and others, over-exploitation
is also leading to aquifer salinization by the mobilization of salt water bodies which
are in contact with the fresh water resources.

The safe yield of all groundwater aquifers of the country can now hardly cover
the municipal and industrial demand; nonetheless these groundwater resources are
over-exploited at a rate of something like 300 MCM/year, mainly used in irrigation
along the highlands.

Even if no groundwater were allowed to be used for irrigation and if groundwater
quality deterioration were stopped immediately, the groundwater stock of Jordan
would not recover. The incurred damage during the last three decades is irreparable,
and irreversible if groundwater bodies are not allowed to recover by a policy aiming
at extracting amounts less than the safe yields.

Most landfill sites in Jordan were chosen at times when the potential environmental impacts of solid waste disposal were locally not fully taken into consideration or not well understood. Therefore, pollution-preventive measures in the design
of the landfill areas or disposal practices were not even thought of, let alone preventive measures such as separation of solid waste constituents in food, glass, plastic
and other remains and their recycling.

Pollution caused by solid waste disposal and/or its pollution potentials in Jordan
shows that the approach and management of municipal solid waste disposal must
undergo radical changes, including separation of wastes, reuse and recycling. Better
protection precautions for surface and groundwater resources by a set of sophisticated measures should be applied on the disposal site before it is put into operation.
Provision for sophisticated leak detection and reduction are of great importance for
the protection of the water resources and for sustaining the environment.

The choices for increasing water resources within Jordan are limited to sea-water
desalinization at Aqaba and treated waste water reuse. The first choice is very

expensive but can be accommodated within the economy of the country when that water is used for municipal supply. The second choice has been discussed above and has proven to be a real non-conventional source of irrigation water. Another alternative to be considered is importing water from other countries. A feasibility study of the Euphrates River was carried out, but no further action was taken because of the riparian rights of that river and the high cost of implementing the project.

Another proposal, advanced by former Turkish governments, is the construction of two pipelines to supply the Middle Eastern countries in Asia and the Gulf States with water from two rivers in Turkey. The project, "Peace Pipeline," would have cost approximately $21 billion. The name of the project suggests the connection between peace and integrated development in the area. This project will have a good chance of implementation after certain political advances are made and if it proves economically feasible. But as a matter of fact, it is somehow surprising that Turkey suggested that proposal but at the same time it reduced the natural flows of the Tigris and Euphrates to Syria and Iraq. It is more logical to leave more water in these two rivers for the use of other countries as suggested in the peace pipeline and this could create new opportunities for Jordan.

At the regional level also, there might be prospects that have not been thought about in the past. First, due to the decreasing of Israeli dependence on Jordan River resources, Jordan can explore the possibility of increasing its water share from the Jordan River up to the level of its share in the Johnston agreement. This could strengthen the peace process and cooperation between the two countries by alleviating the water shortage in Jordan.

Also, as mentioned earlier regarding the developments in Syria, Jordan could explore finding a formula that guarantees the realization of its water share from the Yarmouk River in the final settlement of the Syrian crises by benefiting from its excellent relations with the United States, the European Union, and Russia.

Only expensive projects to utilize water resources can postpone Jordan's crisis a little longer. But even limiting agricultural uses at their present water-consumption rate, allowing domestic demand to cover only the natural increase in population without any rise in living standards or in per-capita consumption and letting planned industries obtain minimal amounts of their needed water, using all the available resources and developing them to safe yield limits, have lead during the last three decades to over-exploitation of almost all aquifers and damage to some of them.

The present severe shortage of water resources and the expected sharpening of demand should give rise to water policies involving more efficient conservation systems rather than the traditional search for new resources. The challenge facing us is to develop and introduce the necessary technologies for water and waste water systems. The increase in population makes this challenge more difficult. The traditional policy of developing new inland resources to satisfy needs is, in the case of Jordan, already exhausted. Now is the time to formulate new policies and change management strategies. Investment in leakage detection and in maintenance and raising water use efficiency in irrigation are more economical ways to increase the water supply. Water leaking from pipes represents a great loss since, although it has been collected, purified, pumped and distributed, it does not reach the consumer to pay for it.

It is now necessary for waste water treatment and reuse to become an integral part of water services. Although waste water is polluted, proper treatment can make its application in irrigation quite safe. It also has advantages over fresh water: waste water contains the nutrients necessary to support plant growth.

The government of Jordan pays the capital cost of all the large irrigation projects. Although it is expected that farmers would irrigate their crops more efficiently if irrigation water prices reflected the actual cost, subsidizing irrigation water is still government policy. Pricing this water artificially low has led to the inability to satisfy the demand. Users of fossil-water resources for irrigation in Azraq, Dhuleil, Disi and other areas pay only for the pumping costs, but not for exhausting these non-renewable national resources. Current practice in this area will certainly lead to the depletion and the loss of the nation's future water and food security. Paying a certain cost now might lead to saving and conserving at least part of the water and may also lead to reconsideration of the economic feasibility of projects.

In the coming decade high-cost projects, environmental hazards and tightened budgets will make large water projects unattractive and difficult to implement. Therefore, policy-makers should add changes in strategy to their policies, such as lowering the demand for and increase the efficiency of water transportation and use instead of increasing the supply.

The increasing demand for water, as a result of population growth and improvements in the standard of living, is gradually leading to competition for the water resources. Projects of additional supplies are becoming more and more expensive and very scarce because of the unavailability of additional resources. Such a situation is expected to gradually lead to the economic consideration of water supply and allocation practices.

In the past four decades social and political issues determined the water use allocations in the country. But the scarcity of water and the expenses of allocating new resources require new thinking and new management procedures.

Water allocations for certain agricultural activities may have to be curtailed, which may in turn cause difficult socio-economic and political problems (more expenditure on foreign currency, increasing unemployment, less food production and eventually social unrest).

The increasing demand for water and the competition among water use sectors will make the present management and development of the water sector through the policy of project-by-project, area-by-area or user-group by user-group planning insufficient. Therefore, the country has to develop a water strategy with adequate dynamic instruments in it to enable comprehensive planning.

Ad hoc decisions in water management are never appropriate because water development and allocation decisions have generally long-term effects on the human activities relying on the water, on the socio-economic and socio-cultural state of the population and on other environmental elements. The only guarantee to consider all these aspects in water management has been the development of a dynamic, comprehensive water strategy, which was based on economic efficiency objectives, while taking into consideration the socio-economic and socio-political components.

The change to an efficient water economy will not be an easy task. But such a change should start and continue. The technologies are available. Therefore, allocating more funds to improving the efficiency of water systems will make unnecessary some expensive, environmentally unsound projects, such as some of those carried out during the 1980s and 1990s.

Economic restructuring from irrigated agricultural to industrial is the way of the future for a country like Jordan, poor in water resources but rich in its talents and its people and enjoying security.

8.5   Pollution Control, Management and Cost

Like all other material resources, water in Jordan should be considered as a resource having the value of a common good. An adequate water supply in terms of quantity and quality offers to people the facilities to enjoy health and a pleasant life.

A clean water supply and healthy water resources base are like a satisfactory level of nutrition, a good desired by all human beings. Thus societies endeavor to promote measures leading to a clean water supply and water resources base. The competition of the different use sectors for this scarce resource; water, in Jordan puts more and more pressure towards a compromise of tolerable water pollution situation and economic development especially because the other use sectors, irrigation and industry produce other goods desired by the society.

Hence, it seems that the resolution of water pollution and resources depletion problems is in itself a resolution of a conflict of interests. Therefore, measures to reduce water pollution levels, to stop resources depletion and to reverse it will positively affect the resources as a common good and hence the society at large, but at the same time these measures will impose costs on the pollution producers and lead to curtailing certain economic activities (irrigation in the highlands). This cost is then effectively carried by the society itself especially in a country like Jordan where water rights (use or property) and pollution control laws even if they were developed are not necessarily enforced. It is worth mentioning that the legitimate uses of property are, in the case of water, not yet well defined, although shouldn't this be one of the main functions of the legal system? The fact that the ownership of a property is essentially the ownership of the rights to some, but not all the services that property can offer has not yet been rooted in the thinking of the population of Jordan.

Also the absence of carefully defined responsibilities for a functioning water resources agency enabling it to maintain quality and water property rights lead to unlimited use (e.g., in irrigation, or private wells of industry) and to eventual conflicts between different types of use. At the same time unrestricted access (in quantity and quality) to water resources by certain groups (farmers, industrialists) leads to over-exploitation and damage of resources as illustrated in the section on pollution.

At this point a forceful type of management should be implemented in order to limit the damage and restore the resources quantitatively and qualitatively.

In order not to risk irreparable damage, early preventive actions are required, even in advance of clear-cut scientific evidence of teh relations between water extraction, disposal of waste water or solid wastes on the one hand, and depletion of water resources and their quality degradation on the other.

Therefore, in the case of Jordan, the policy of viewing water degradation issues as being less important, and giving greater priority to economic development should be rapidly abandoned in the interest of present and future generations and water resources sustainability. Here, it should be emphasized that water resources cannot always be regarded as an externality in the production process, but as an inherent part of the production and consumption process and that sustainability principles must address that adequately, especially in what concerns water pricing and pollution charges. If, for whatever reason, whether institutional, administrative or tribal, or because of influential groups of beneficiaries … etc. water resources have to bear a zero price or a symbolic price, then other regulations and behavioral rules have to be superimposed to reduce and counteract the social cost versus private cost discrepancies.

Such regulations and behavioral rules may contain appropriate environmental standards, taxes on pollution and other policy instruments.

Intergeneration equities in quantitative and qualitative terms have to be one of the major instruments which should govern water developmental policies. They should not be allowed to irreversibly pollute or over-exploit a water resource unless the revenues of that are invested for the benefits of future generations to substitute the damage or the depletion of resources, especially those resources which are non-renewable or fossil.

Protection, enhancement and restoration of water quality and abatement of water pollution should be the major component of any water development program because preservation of the resources base is very vital for sustainable development. This is especially valid when knowing that water pollution and resources depletion abatement costs are far less than the damage cost.

For that it is imperative to enable policy makers to judge the effects of depletion and degradation of water quality by valuating pollution damage based on the magnitude of the damage in physical terms (m^3 of water) and an agreed upon means of converting that into a common unit of measurement in monetary terms.

A water polluter is likely to ignore the consequences of his activities for others if those affected groups of the society are not quite aware of the magnitude and impact of pollution or if they for any reason (being employed by the polluter, or paid by him, or socially linked to him) suppress their sufferings and the damages their water resources are experiencing. Even governmental agencies try to ignore the consequences of their activities such as depleting aquifers in the Jordan Valley (flowing wells of brackish water) damaging the resources base – if the affected groups do not

Table 8.2 Water resources problems in Jordan

Inadequate resources	Geography	Cost recover	Population	Water quality deterioration	Water supply network	Illegal connection
1. Climate of Jordan arid to semi-arid.	1. Cities and towns lay generally in the high mountainous areas.	1. Low water prices, subsidies to consumers.	1. High population growth refugees.	Population, animal and plant life and the environment	Inadequate, bad material, bad installations maintenance and repair	
2. Water use by other riparian countries.	2. High cost of pumping, maintenance and repair	2. Extended water supply networks.	2. Higher living standards.			
3. Salinity and inappropriate quality for certain uses.		3. Land use planning irrelevant	3. Remoteness of population centers from water resources			
4. Use fresh water for irrigation.						
Affected area	Whole country	Highlands and Badia	Whole country	Amman, Zarqa, Azraq, Jafr, J. Valley, Dhuleil	Whole country	Whole country

oppose them or if there is no directly affected group. Against such damage only NGOs can raise their voices in the interest of the society and future generations.

Sewerage treatment plants should be designed to accommodate organic and hydraulic loads for the expected population and discharges after 30 years. In Jordan all the waste water treatment plants designed and implemented in the second half of the 1980s were overloaded hydraulically or organically after a few years (Table 8.2). In all the standards this indicates incompetence, short-sightedness and the wish to under design to make projects attractive to the government. Those same designers and implementers of such projects thus keep themselves busy because they have to redesign and re-implement after a few years to accommodate the additional loads. (It is a continuous flow of income to them.)

Groundwater extraction should be limited to aquifer safe yields; it should have no adverse effects on the groundwater quality or quantity (Table 8.3).

If non-renewable groundwater resources have to be exploited (mined) then the revenues of their use (even that of domestic uses) should be able to cover their substitution or they should be invested to enable future generations to substitute them from other resources or by using other technologies (Disi, Azraq, etc.).

Preventive protection measures against groundwater pollution by the infiltration of lower quality water, such as treated or untreated waste water, irrigation return

Table 8.3 Water quality problems in Jordan

Type of pollution vs. its effects	Deterioration of water quality as result of over pumping	Deterioration waste water quality due to low water consumption	Inadequate waste water treatment, biocides and pharma residues	Waste water low coverage	Irrigation return flows	Solid waste disposal	Industrial waste water
Affected area	Overall north, partly central and southern Jordan	Whole country	Whole country	Overall where cesspools are in use	All over where irrigation is practiced	All over where solid wastes disposal is practiced	Amman, Zarqa, Balqa, Aqaba, Irbid etc.
Affected environmental element	Ground water and surface water	Water, soil, plants and food	Water soils, and food	Ground and surface water	Ground and surface water.	Soils, surface and groundwater, air quality.	Surface and ground water, plants, soils
Causes	Inadequate water quality, deteriorating surface and ground water	Polluted irrigation water, soils, ground and surface water	Deteriorating soils, surface and ground water qualities and human and animal health.	Human and animal health, soils, surface and ground water qualities and quantities.	Deteriorating soils, surface and ground water qualities and lower land productivity.	Increasing salinity of soils, ground and surface water and trace elements and pharma residues	Contamination of surface and ground water with trace elements, salinity industrial chemicals.

flows and others should be well designed and implemented before initiating projects. This requires that cross-sectoral issues related to water extraction, use and discharge such as land use, industrial and agricultural activities should be part of the overall approach to water management. In the case of Jordan, the repair of damage to groundwater proved to be inefficient, costly and impossible in some regions (Dhuleil, Jafr, Azraq, Wadi Dhuleil and more recently Za'atari and Agib as a result of Syrian refugees).

Reference

Salameh E, Udluft P (2001) "Towards a water strategy for Jordan" Hydrogeologie und Uwelt, Wurzburg University, Germany